Whitestein Series in Software Agent Technologies and Autonomic Computing

Series Editors:
Monique Calisti (Editor-in-Chief)
Marius Walliser
Stefan Brantschen
Marc Herbstritt

The Whitestein Series in Software Agent Technologies and Autonomic Computing reports new developments in agent-based software technologies and agent-oriented software engineering methodologies, with particular emphasis on applications in the area of autonomic computing & communications.

The spectrum of the series includes research monographs, high quality notes resulting from research and industrial projects, outstanding Ph.D. theses, and the proceedings of carefully selected conferences. The series is targeted at promoting advanced research and facilitating know-how transfer to industrial use.

About Whitestein Technologies

Whitestein Technologies is a leading innovator in the area of software agent technologies and autonomic computing & communications. Whitestein Technologies' offering includes advanced products, solutions, and services for various applications and industries, as well as a comprehensive middleware for the development and operation of autonomous, self-managing, and self-organizing systems and networks.
Whitestein Technologies' customers and partners include innovative global enterprises, service providers, and system integrators, as well as universities, technology labs, and other research institutions.

www.whitestein.com

Advanced Agent-Based Environmental Management Systems

Ulises Cortés
Manel Poch
Editors

Birkhäuser
Basel · Boston · Berlin

Editors:

Ulises Cortés
Departamento Software
Universitat Polytècnica de Catalunya
Jordi Girona 1
08034 Barcelona
Spain
e-mail: ia@lsi.upc.edu

Manel Poch
Laboratori d'Enginyeria Química i Ambiental
(LEQUIA)
Facultat de Ciències
Universitat de Girona
Campus de Montilivi
17071 Girona
Spain
e-mail: manel@lequia.udg.es

2000 Mathematical Subject Classification: 68-06, 68U35, 68N19, 68T05, 68T35, 68T27
1998 ACM Computing Classification: I.2.11[Distributed Artificial Intelligence]: Intelligent
agents; I.2.11[Distributed Artificial Intelligence]: Multiagent systems; H.3.5 [Information
Storage and Retrieval]: Online Information Systems: Web-based services; H.3.4 [Information
Storage and Retrieval]: Systems and Software: Distributed Systems

Library of Congress Control Number: 2008941124

Bibliographic information published by Die Deutsche Bibliothek
Die Deutsche Bibliothek lists this publication in the Deutsche Nationalbibliografie;
detailed bibliographic data is available in the Internet at <http://dnb.ddb.de>.

ISBN 978-3-7643-8897-3 Birkhäuser Verlag AG, Basel – Boston – Berlin

© 2009 Birkhäuser Verlag, P.O. Box 133, CH-4010 Basel, Switzerland
Part of Springer Science+Business Media
Printed on acid-free paper produced from chlorine-free pulp. TCF ∞

ISBN 978-3-7643-8897-3   ISBN 978-3-7643-8900-0   (eBook)

9 8 7 6 5 4 3 2 1   www.birkhauser.ch

Contents

Whitestein Series in Software Agent Technologies and Autonomic Computing, 1–4
© 2009 Birkhäuser Verlag Basel/Switzerland

Preamble

Ulises Cortés and Manel Poch

Our environment is precious. Our quality of life depends on the optimal management of its resources. As we know nowadays, it is a very complex ecosystem, being the result of a wide set of species and elements that interact in a complex net. But the environment is fragile. It is well-known that the set of relationships between the different subsystems may be altered by human actions. These actions affect the biodiversity. Thus, government, industry and the public must act to avoid further damages. We must be capable of developing and applying an integrated approach to affront the problem(s) in the most efficient way.

But this is not easy. Ecosystems are complex and not well-known in their dynamics. Moreover, they present diverse scales (spatial and temporal). Problems arise when the quantity of available information is huge and nonuniform, coming from many different sources often characterized by great uncertainty and often their quality cannot be stated in advance.

Thus, government, industry, and the public call for integrated environmental management systems capable of supplying all parties with validated, accurate and timely information. An effective protection of our environment is largely dependent on the quality of the available information used to make an appropriated decision.

The *near real-time* constraint on the needed answer reveals two critical problems in delivering such tasks: the low quality or absence of data, and the changing conditions over a long period. Another associated issue is the dynamical nature of the problem.

Computers are central in contemporary environmental protection in tasks such as monitoring, data analysis, communication, information storage and retrieval, so it has been natural to try to integrate and enhance all these tasks with knowledge-based techniques from Artificial Intelligence [3, 2].

Application of Information and Computation Technology (ICT) to the environment is a very broad topic that covers the range from satellite observation to miniature sensors, from flood prediction to noise measurement. Due to the nature of environmental models, *i.e.* their complexity, nonlinearity, dynamics and spatially distributed nature, many of the classical methods of error analysis are difficult to apply as they require differentiable models.

Environmental problems require a new approach to decision support for two fundamental reasons:

1. It is impossible to solve the inverse problem *directly* due to the complexity of the systems.
2. It is impossible to solve decision problems unequivocally due to the complexities and changing nature of the decision making process itself [4].

With these problems in mind, important work has been made to develop Environmental Decision Support Systems (EDSSs) as a tool to improve complexity management, becoming a reference in environmental problems solution [6]. The majority of developed EDSSs are based on the traditional rule-based approach; this was a step ahead but it was not enough. Since the 1990s, agent-based approaches appear as a promising alternative. Agents are an approach to building a wide range of environmental applications (see for example [1]). Agents are autonomous problem-solving entities that are able to flexibly solve problems in complex, dynamic environments.

Agent-based approaches introduce a powerful metaphor, having the flexible autonomous action required to adapt to the changing conditions. Agent technology represents an alternative worth to be explored as there are many phenomena in environmental systems that can be characterized by interaction between agents and environment. Environmental applications require agents to be able to interact with numerous other agents in order to achieve their goals [5].

This book presents a collection of papers reflecting the interest and impact of agent technology on the definition and development of Environmental Decision Support Systems.

Outline of the Book. We have selected six papers that in our understanding illustrate the growing interest in the use of agent technology to solve complex problems in the environmental domain:

- Chapter 1. Agents as a Decision Support Tool in Environmental Processes: The State of the Art *(M. Aulinas, C. Turon & M. Sànchez-Marrè)*.
- Chapter 2. Deliberation Over the Safety of Industrial Wastewater Discharges into Wastewater Treatment Plants *(P. Tolchinsky, M. Aulinas, M. Poch & U. Cortés)*.
- Chapter 3. OSM: A Multi-Agent System for Modeling and Monitoring the Evolution of Oil Slicks in Open Oceans *(J.M. Corchado, A. Mata & S. Rodríguez)*.
- Chapter 4. Designing an Information System for the Preservation of Insular Tropical Environment in Reunion Island Integration of Databases, Knowledge Bases and Multi-Agent Systems by using Web Services *(N. Conruyt, D. Sebastien, D. Payet, Y. Geynet, D. Caron, D. Grosser & R. Courdier)*.
- Chapter 5. A Methodology for Developing Environmental Information Systems with Software Agents *(I.N. Athanasiadis & P. A. Mitkas)*.
- Chapter 6. Environmental-knowledge Management for Cognitive Agents *(L. Ceccaroni, A. Simón-Cuevas, A. Rosete-Suarez & M. Moreno-Espino)*.

The first paper presents a state of the art of agent-based applications in environmental management. A wide revision of agent-based tools is presented showing the variety of approaches and problems considered.

Papers 2 and 3 consider two specific cases on *how* to manage discharges in the environment, e.g. technological depollution devices such as wastewater treatment plants and oil spills in open oceans. These are important and actual examples of problems in the interaction between human activities and the environment. Different reasoning approaches are used in both cases to offer a solution.

Paper 4 and 5 consider the design of agent-based environmental information systems in two very different contexts: (1) The fragile ecosystem of La Reunion Island and (2) in the Mediterranean context. Information systems have become the backbone of all kinds of organizations today. These papers are good examples of the next generation of mainstream information systems, which we might term *active computing*. Organizations dealing with the environment and its related problems need agent-based approaches, because of their rich representational capabilities that allow for more faithful and flexible treatments of complex fluxes of information.

Finally, paper 6 considers the knowledge representation problem. The authors propose the use of cognitive agents as a metaphor in the design of knowledge-based systems. In particular, they suggest the use of *Concept Maps* as a way to deal with the representation.

In order to be useful to the widest range of stakeholders involved in environmental studies and management, from researchers to policy makers, we decided that each paper should present an integral study of each considered problem. We have asked the authors for considerably extended versions of the originally selected papers, to assure durable relevance of the information presented in this book.

We are sure that the diffusion of environmental innovations is one way towards a more sustainable development. Our aim when editing this book is to present some innovative and successful approaches that use agent-based technologies that give answers to real environmental problems. The papers presented in this book show that the work made in this area leads to useful tools that improve our capacity for more optimal management of the impact of human activities on the environment.

We hope that the book serves as a tool for an audience, that includes postgraduate students, practitioners in consulting engineering, decision makers in national and international regulatory bodies and other researchers, to develop new scenarios for the future use of agent-based technologies, and therefore, indicating ways towards the collective construction of a sustainable future.

References

1. I.N. Athanasiadis, *A review of agent-based systems applied in environmental informatics*, Proc. of Int'l Congress on Modelling and Simulation (MODSIM), 2005.

2. S.H. Chen, A.J. Jakeman, and J.P. Norton, *Artificial Intelligence techniques: An introduction to their use for modeling environmental systems*, Mathematics and Computers in Simulation **78** (2008), 370–400.

3. U. Cortés, M. Sànchez-Marrè, L. Ceccaroni, I. R-Roda, and M. Poch, *Artificial Intelligence and Environmental Decision Support Systems*, Applied Intelligence **13** (2004), no. 1, 77–91.

4. K. Fedra, *Environmental Decision Support Systems: A conceptual framework and application examples*, Ph.D. thesis, Facultè des sciences de l'Universitè de Genéve, 2000.

5. M. Luck, P. McBurney, O. Shehory, and S. Wilmott, *Agent Technology: Computing as Interaction (A Roadmap for Agent Based Computing)*, AgentLink, 2005.

6. M. Poch, J. Comas, I. Rodríguez-Roda, M. Sànchez-Marrè, and U. Cortés, *Designing and building real environmental decision support systems*, Environmental Modelling and Software **19** (2004), no. 9, 857–873.

Ulises Cortés
Knowledge Engineering and Machine Learning group
Dept. Llenguatges i Sistemes Informàtics
Universitat Politècnica de Catalunya
Jordi Girona 1-3
Barcelona 08034
Spain
e-mail: `ia@lsi.upc.edu`

Manel Poch
Laboratory of Chemical and Environmental Engineering
University of Girona
Campus de Montilivi s/n
Girona 17071
Spain
e-mail: `manuel.poch@udg.edu`

Whitestein Series in Software Agent Technologies and Autonomic Computing, 5–35
© 2009 Birkhäuser Verlag Basel/Switzerland

Agents as a Decision Support Tool in Environmental Processes: The State of the Art

Montse Aulinas, Clàudia Turon and Miquel Sànchez-Marrè

Abstract. Agent-based systems have become an important area of research since the 1990s. They have been applied to a range of domains that are intrinsically complex. Among these, environmental problems are of special concern, given their ample affectation to our societies and everyday quality of life. This report provides a review of agent-based systems applied to environmental problems of diverse nature. The usefulness of Multi-Agent Systems (MASs) to model complex systems that embed multiple and dynamic interactions, such as in environmental processes, is revealed.

Keywords. Agent-Based Modeling, Environmental Processes, Multi-Agent Systems, Natural-Resources Management.

1. Introduction

The constituent parts of the environment (*i.e.* different life forms, energy and material resources, and the atmosphere) interact with each other. As an example, changes in biosphere composition affect the atmosphere composition. For our concern, more important are the effects of human activity on the environment, and the consequences of these affects on human well-being.

All environmental problems are essentially related to the use and distribution of resources, affecting water, air and soil quantity and quality. Environmental problems can be categorized in a number of ways. However, *resources overexploitation* and *environmental pollution* is amongst the most serious of existing category of problems. The growth of population and economic wealth, together with the increase of several processes, such as urbanisation and industrialization, has lead to a high consumption of natural resources and consequently, negative effects on the sustainability of the environmental quality have risen.

On the one hand, some forms of pollution can disrupt complex biogeochemical cycles and on the other, pollution brings significant social and economic consequences. As follows, we portray briefly the three main categories of environmental pollution: water, air and soil pollution. Their related problems are of special concern.

Water pollution (in oceans, rivers, lakes, aquifers, *etc.*) is any chemical, physical or biological change in the quality of water that has a harmful effect on any living organism that drinks or uses or lives (in) it. Water pollution sources are often classified as point and non point sources. Point sources discharge pollutants at specific locations through pipelines or sewers into the water bodies (*e.g.* industries, sewage treatment plants, underground mines, oil wells, oil tankers, *etc.*). Non point sources are sources that cannot be traced to a single site of discharge (*e.g.* acid deposition from the air, traffic, pollutants that are spread through rivers, agriculture runoff, *etc.*).

Air pollution supposes the introduction into the atmosphere of chemicals, particulate matter, or biological materials that cause harm or discomfort to humans or other living organisms, and/or damages the environment. Although air pollution is often identified with major stationary sources (*i.e.* industries), the greatest source of emissions is produced by mobile sources (*i.e.* automobiles).

Finally, soil pollution is caused by the presence of man-made chemicals and other alteration in the natural soil environment. This type of pollution commonly is originated due to the application of pesticides, the percolation of contaminated surface water to subsurface strata, oil and fuel dumping, leaching of wastes from landfills or direct discharge of industrial wastes to the soil, *etc.* It affects directly the habitat in which biodiversity is embedded and any other natural and/or human land uses.

Environments have some capacity to absorb and neutralize many substances (resilience), so a distinction is often made between *pollution*, involving harmful effects, and *contamination*, the presence of a substance in the environment below the damage threshold. This distinction is very clear in theory, but sometimes very difficult to establish in practice. In fact, environmental problems at all scales - from the merely local to those with long-term global significance - raise certain fundamental issues which make their resolution difficult and controversial. Some recurrent issues, many of them interrelated, include the following:

- Environmental problems are *multidisciplinary* by nature. As a consequence, in most environmental management situations, a single expert who can solve the problem entirely does not exist. Different opinions about the causes, consequences and possible solutions for the problem exist. Thus, conflict is inherent when trying to solve environmental problems due to the multiplicity of views and interests involved.
- Environmental problems are often characterized by great *uncertainty*. The complexity of environmental systems means that our understanding of the

human impact upon it is very partial, and accurate prediction is often impossible. Collected environmental information is often imprecise, uncertain or erroneous. As knowledge advances, uncertainties are reduced, but they can rarely be eliminated.

- Environmental problems involve strong *spatial* and *temporal distribution*. The multiplicity of scales has been traditionally associated with distinct spatial scales (*i.e.* local, regional, global), each associated with specific timescales. The irregular distribution of environmental problems in time and space make difficult to well define the interactions among these scales.
- Environmental problems are hard to model and understand. Environmental problems, as well as environmental systems, are *dynamic* in nature, and therefore deep models of their behavior are difficult to reproduce.

The experts' reasoning about environmental problems and decision making about suitable solutions is understood as manipulating high amount of specific data, mathematical models of the real situation, simulations, *etc.* In case of *inaccessibility, incompleteness,* or *incorrectness* of data as well as in other situations with high degree of uncertainty, experts still are able to make decisions. However they need to understand, in a limited time, chemical, physical and biological processes in relation to socioeconomic conditions and applicable legislative framework. The high complexity of environmental problems, characterized by the aforementioned most frequent issues, has lead to the use of knowledge-based decision support tools in decision processes.

2. Environmental Decision Support Tools and Agent-Based Paradigm

Over the last few decades, mathematical/statistical models, numerical algorithms and computer simulations models have been used as an appropriate means to gain insight into environmental management problems and provide useful information to decision makers. To this end, a wide set of scientific techniques has been applied to environmental management problems for a long time and with good results. The effort to integrate new tools to deal with more complex systems has led to the development of so-called Environmental Decision Support Systems (EDSSs) ([31], [71]).

EDSSs have generated high expectations as tools to tackle high complex problems. The range of environmental problems to which EDSSs have been applied is wide and varied, with water management at or near the top, followed by aspects of risk assessment and forest management. Equally varied are the tasks to which EDSSs have been applied, ranging from monitoring and data storage to prediction, decision analysis, control planning, remediation, management, and communication with society. Environmental issues belong to a set of critical domains where wrong management decisions may have disastrous social, economic and ecological consequences. Decision-making performed by EDSSs should be collaborative, not adversarial, and decision makers must inform and involve those who must live with the decisions. EDSS should be not only an efficient mechanism to find an optimal or sub-optimal solution, given any set of whimsical preferences, but also a mechanism to make the entire process more open and transparent.

According to Fox and Das [27], a decision support system is a computer system that assists decision makers in choosing between alternative beliefs or actions by applying knowledge about the decision domain to arrive at recommendations for the various options. It incorporates an explicit decision procedure based on a set of theoretical principles that justify the "rationality" of this procedure.

Intelligent Environmental Decision Support Systems (IEDSS) are intelligent information systems that reduce the time in which decisions are made in an environmental domain, and improve the consistency and quality of those decisions, by integrating several types of information and knowledge [67].

IEDSSs are tools designed to cope with the multidisciplinary nature and high complexity of environmental problems. The main advantages of using IEDSS to solve environmental problems rely on [71]:

1. The ability to acquire, represent and structure the knowledge.
2. The possibility to separate the data from the models.
3. The ability to deal with spatial data (incorporating GIS tools, for example).
4. The ability to provide expert knowledge (incorporating specific knowledge bases).
5. The ability to be used effectively for diagnosis, planning, management and optimization.
6. The ability to assist the user during problem formulation and selecting the solution methods.

Thus IEDSSs could be defined as systems using a combination of models, analytical techniques, and information retrieval to help develop and evaluate appropriate alternatives ([77], [1], [78]); and such systems focus on strategic decisions and not operational ones. More specifically, decision support systems should contribute to reducing the uncertainty faced by managers when they need to make decisions regarding future options [30]. Distributed decision making suits problems where the complexity prevents an individual decision maker from conceptualizing, or otherwise dealing with the entire problem ([14], [16]).

The use of AI tools and models provides direct access to expertise, and their flexibility makes them capable of supporting learning and decision making processes [67]. This confers on IEDSSs the ability to confront complex problems in which the experience of experts provides valuable help for finding a solution to the problem. It also provides ways to accelerate identification of the problem and to focus the attention of decision-makers on its evaluation.

Most developed IEDSS are based on traditional artificial intelligence approaches. These approaches are clearly bounded in the way and reliability they can solve the aforementioned environmental issues. However, since the 90s, the agent-based paradigm has emerged as a potential tool to deal with the interaction of humans and ecosystems, citizens and stakeholders, as they are tools designed to cope with the multidisciplinary and distributed nature and high complexity of environmental problems. An intelligent agent could be defined as any autonomous entity that is capable of perceiving its environment and carrying out goal-directed action ([72], [86]). *Agent-based approaches* have introduced both a powerful metaphor and a group of technologies in the field of IEDSS, giving support to the management of environmental problems, mainly of those concerning the management of renewable resources (*e.g.* water management, biodiversity management, forest management, erosion and soil management, *etc.*). These problems represent typical dynamic and unpredictable multi-agent domains, where flexible autonomous action is required to adapt to changing conditions. The need to cope with dynamic and emergent situations requires application components to interact in more flexible ways. The characterization in terms of *agents* has proven to be a most natural abstraction to many real world problems, having convinced researchers and developers in a wide variety of domains (*e.g.* [59], [35], [4]) of the great potential of multi-agent solutions.

Multi-agent systems are built describing numerous agents with a different degree of complexity, according the information and knowledge available. The complexity and degree of knowledge can be improved without the necessity to modify all the system (they are intended to be systems with high modularity and scalability). The system can be extended with the addition of new types of agents and adding new capabilities into the implicit ontology without a central control, by just publishing it as part of agent self-description [43].

Briefly, for *modular, decentralized, changeable, ill-structured* and *complex system*, software intelligent agents are really appropriate [61].

3. State of the Art of Agent-based Applications in Environmental Management

In [4] a review of various published applications are considered. The review is done from both agent-oriented software modeling and implementation perspectives. Athanasiadis remarks that the applications can use agent-based approaches and methods, either as a metaphor for software design or as an abstraction for software development. The applications (an overall of 23 dating from 1996 to 2004) are grouped in three categories to ease their presentation:

1. *Environmental information and data management* (Environmental Data Management Systems (EDMSs)). In most of environmental problems available data and information is characterized by the attributes mentioned in §1: uncertain, imprecise, incorrect, and spatially distributed. EDMSs are needed to tackle with this kind of information. EDMSs are aimed at managing, integrating or distributing environmental data.

2. *Decision support in environmental problems* (Environmental Decision Support Systems (EDSSs)). Most of the applications in this category use agent methodologies and technologies in a way to make the decision-making distributed and shared between the different experts and stakeholders involved in specific environmental problems.

3. *Simulation of environmental or ecological systems and processes* (Environmental Simulation Systems (ESSs)). Agent-based ESSs use agents as the structuring blocks for modeling processes and interactions. The growing interest in this technique is due to the possibility to incorporate almost directly and intuitively the behavior observed in the real world by means of a computational model.

Then, the applications are reported with its main tasks and objectives, the application field, related technologies and principal agent types involved. Next, they are evaluated in terms of their level of software (SW) design and development (from low to upper level design, and from objects to agent-platforms implementation, respectively). In Tables 1–5 (see pp.22–26) we update and rationalize the available agent-based applications in environmental management following partially the criteria used in [4], and continuing the revision from 2005. The classification of applications in one of the three aforementioned categories (*i.e.* EMS, EDSS and ESS) is not always obvious, since the boundaries between the three categories are intertwined and not always clearly discriminated. The overview of applications is presented chronologically ordered (from the oldest to the newest published references). Four columns have been added to better describe the systems reviewed. These columns make reference to:

Software design: From this aspect it is possible to analyze the use of agent-related technologies in software design and modeling. That is, how the agent's concept is used. According to [4] four levels of agent's design complexity can be distinguished:

(1) At the lowest level there are systems that use some agent-alike *entities*.

(2) In the second level the systems are modeled using agents (a model), typically involving UML design.

(3) The third level involves agents for software specification, that is the use of BDI [69], LORA [87] or similar techniques.

(4) In the fourth level the systems adopt a sophisticated agent-oriented software design process as Gaia [88] or Tropos [29].

Software development: From the point of view of software implementation four levels of agent-related technologies can be identified:

(A) Implementation with objects.

(B) Implementation with software agents, typically dealing with FIPA standards (http://www.fipa.org).

(C) Implementation using available agent platforms such as JADE, ZEUS, JACK, *etc.*.

(D) Implementation using an own platform.

Implementation: In this column we refer to the system's implementation phase or stage. That is, if the reviewed systems are in the design phase or at the beginning of the development, partially or fully implemented, in progress, *etc.* Somehow it completes the information give in the 'software development' column.

Validation: In this column information on whether the agent-based system has been or not tested is given. In computer modeling and simulation, validation is the process of determining the degree to which a model or simulation is an accurate representation of real world from the perspective of the intended uses of the model or simulation.

As follows, a brief explanation of each of the applications reviewed is provided. In this review we have only considered those applications related to environmental management issues; other domains such as economics ([20], [34], [80]), telecommunications ([85], [15]), healthcare ([33], [55], [3]), manufacturing ([62], [17]), military support [82], *etc.* sustain the suitability of agent-based applications in complex domains. However, although the intention is to present only those agent-based applications related environmental management issues, some of the applications are either not developed exclusively with agents or they do not deal solely with environmental management applications.

The **DAI-DEPUR** system applies distributed artificial intelligence techniques in a decision support system for supervising a wastewater treatment plant. The processes of the plant are represented by agents, which collaborate in a layered architecture [74]. This supervisory integrated and distributed architecture proposes the integration of several interacting subsystems or agents, and the combination of problem solving capabilities, reasoning as well as learning tasks in a single structure. A real world application was delivered later in ATL-EDAR [75].

In the **EDS** (*Environmental Decision Support*) application an agent community is used for supporting the decision-making process related with environmental assessment, planning, and project evaluation. Specifically, the *EDS* system provides assistance to project developers in the selection of adequate locations of their projects (*e.g.* roads, industries, hospitals, *etc.*), guaranteeing the compliance with the applicable regulations and the existing development plans as well as satisfying the specified project requirements and the fulfilment of applicable regulations according to the location ([44], [45]).

The **SAEM** system (*A Society of Agents in Environmental Monitoring*) proposes the use of robotic agents that collaborate for monitoring and evaluating the pollution on a power plant chimney [76]. Specifically, a simulated application of small flying robotic agent societies (helicopter models) is assigned to go around a chimney in order to sample the pollutant cloud and to send values to a central processing unit which builds a global map. This map is then transformed into an image that holds information about cloud direction, pollutant concentration, *etc.*

allowing decision makers to evaluate and change the burning conditions of the power plant.

In the **ESAT-WMR** system (*Expert System and Agent Technology to Water Mains Rehabilitation*), the agent-based decision support tool reported intents to support a U.K. water company in its water mains rehabilitation decision making processes. A community of collaborative agents models the tasks and interactions of the water company and its associates, and, ultimately, assesses alternative strategies for the pipes network rehabilitation ([23], [24]).

The **IDS-DAP** system (*Intelligent Decision Support System for Differentiated Agricultural Products*) is a DSS applied for the selection of agricultural product penetration strategy. It incorporates distributed multi-criteria analysis models. Concretely, the multi-criteria method UTASTAR is applied to the multi-criteria consumer preferences in order to determine the criteria explaining each of the consumer's choices into consumer agents participating in a particular market research ([49], [48]).

The **FIRMA** project (*Freshwater Integrated Resource Management with Agents*) applies agent-based modeling for the integration of natural, hydrologic, social and economic aspects of freshwater management. A variety of agent-based models has been developed for simulating consumers, suppliers, and government, and their interactions at different scale of aggregation. One of the *FIRMA* test cases has been applied on the Thames River to explore the effects of precipitation and temperature on water availability and household demand [12]. In this case, water consumer agents communicate with each other, sharing perspectives in the form of endorsement [54].

The **SHADOC** system (French acronym for *Hydro-agricultural Simulator describing Organization and Coordination Modes*) uses agents for simulating the behavior of the stakeholders and the farmers involved in the irrigation of Senegal valley [9]. The model constitutes a virtual irrigated system which can already be used as a tool to test hypotheses of social organizations and institutions. This is still a theoretical simulator somewhat specific to the Senegal River Valley even though it has been designed to be able to deal with other contexts.

EDEN-IW (*Environmental Data Exchange Network for Inland Water*) is a system that aims to provide citizens, researchers and other users with existing inland water data, acting as a one-stopshop [25]. *EDEN-IW* exploits the technological infrastructure of *Infosleuth* system ([57], [66]), in which software agents execute data management activities and interpret user queries on a set of distributed and heterogeneous databases. Also, *InfoSleuth* agents collaborate for retrieving data and homogenizing queries, using a common ontology that describes the application field. *EDEN* pilot demonstration enables integrated access via web browser to environmental information resources provided by offices of the connected agencies. The demonstration focuses on information relating to remediation of hazardous waste contamination.

WaWAT (*WasteWater Agent Town*) employs several co-operative agents who make use of case-base reasoning, rule-based reasoning and reactive planning, to support supervision and control of wastewater treatment plants [18]. It uses the WaWO ontology (*Waste Water Ontology*) [19] which provides a set of concepts that can be queried, advertised and used to control agent cooperation.

The **BUSTER** system (*Bremen University Semantic Translator for Enhanced Retrieval*) utilizes ontologies for retrieving information sources and semantic translation into the desired format [56]. This approach can be applied when the information can be accessed by remote systems in order to supplement own data basis. The *BUSTER* approach provides a common interface to heterogeneous information sources in terms of an intelligent information broker. A user can submit a query request to the network of integrated data sources (*e.g.* as shown in a query example sampling information about the land use of a specific site).

Adour is a bargaining model to simulate negotiations between water users in a river basin [81]. A formal computable bargaining model of multilateral negotiations is applied to the Adour Basin case, in the South West of France, with seven agents (three "farmers", two "environmental lobbies", the water manager, the taxpayer) and seven negotiation variables (three individual irrigation quotas, the price of water, the sizes of three dams), in order to negotiate alternatives of water use. A sensibility analysis is conducted to quantify the impact of the negotiation structure (*e.g.* political weights of players, choice of players...) on simulations outcomes. The final aim is to provide a better understanding of the complex interrelations between the various components of the modeled system: preferences of stakeholders over negotiated variables, the role of exogenous (*i.e.* hydraulic and budgetary) constraints in the bargaining game, the consequences of the structure of negotiation (*e.g.* decision rule, players' weights, dimension of the issue space *etc.*) on the bargaining outcome *etc.*

DIAMOND (*DIstributed Architecture for MONitoring and Diagnosis*) adopts an agent-based architecture for distributed monitoring and diagnosis [2]. Industrial diagnostic systems aim at anticipating the occurrence of failures or, should failures have occurred, at detecting them and identifying their cause. *DIAMOND* will be demonstrated for monitoring of the water-steam cycle of a coal fire power plant, and for integrating a diagnostic system with an existing process control network.

The **MAGIC** system (*Multi-Agents-based Diagnostic Data Acquisition and Management in Complex Systems*) was created with the same purpose as *DIAMOND*. Even if it is not targeted only for environmental applications, its objective is to develop a flexible multi-agent architecture for the diagnosis of progressively created faults in complex systems, by adopting different diagnostic methods in parallel. *MAGIC* has been demonstrated in an automatic industrial control application [37].

A quite similar system that uses software agents for accessing environmental data is **NZDIS** (*New Zealand Distributed Information System*). *NZDIS* ([21],

[68]) has been designed for managing environmental meta-data in order to service queries to heterogeneous data sources. *NZDIS* software agents are used for submitting queries to environmental databases in a seamless way. Agents receive and reply to requests for services and information by means of a high level declarative agent communication language, whose message contents may be expressed in terms of formal ontologies that describe the vocabularies of various domains.

The **D-NEMO** experimental prototype, installed in the Athens Air Quality Monitoring Network, uses agents for the management of urban air pollution [36]. *D-NEMO* agents incorporate classification and regression decision trees, case based reasoning and artificial neural networks for forecasting collaboratively air pollution episodes.

The **RAID** system (*Rilevamento dati Ambientali con Interfaccia DECT*) deals with pollution monitoring and control in indoor environments. *RAID* exploits the general architecture of Kaleidoscope that uses "entities" for the dynamic integration of sensors [52]. The system is based on innovative sensors and wireless communication. It includes a knowledge-based supervisor aimed at identifying pollutant sources.

AqEcAA (*Aquatic Ecosystem Simulation with Adaptive Agents*) presents a conceptual framework simulating the aquatic food web and species interactions by using adaptive agents [70]. It provides a realistic framework for ecosystem simulation, evolving ecosystem structures and behaviors by emerging, submerging, interacting and evolving ecological entities.

the **CATCHSCAPE** system [13] deals with the irrigation of northern Thailand, using agents for representing all entities related with the hydrologic basin. Agents incorporate models for the determination of aquatic reservoirs with respect to future changes in drought conditions and changes in commodity prices, and farmer behavior.

The **SINUSE** application [26] employs agents to model the Kairouan water basin. *SINUSE* agent-based system investigates the consequences of human behavior in the availability of aquatic resources by simulating physical and socioeconomic interactions on a free access water table. *SINUSE* is considered as a first step in the use of MASs for groundwater studies, and it has proved the relevance of taking local and non-economic interaction into account in the case of the Kairouan water table.

The **STAU-Wien** application (*City-Suburb Relations and Development in the Vienna Region*) aims to study the urban growth of Vienna city and its suburbs. The objective of this work is to simulate prior and future landscape transition processes for the suburban region in the surroundings of Vienna, Austria. A spatial agent model is used for stimulating regional migration and allocation decisions of households and commercial enterprises [41].

The multi-agent model **GEMACE** (*Multi-Agent Model to Simulate Agricultural and Hunting Management of the Camargue and its Effects*) simulates the interactions between hunters, farmers and duck population of a habitat. The system investigates the correlations between human activities and the environment and their impacts to the land use and the population of ducks [47].

The **FSEP** project (*Forecast Streamlining and Enhancement Project*) is being developed in the Australian Bureau of Meteorology, and uses agents for detecting and using data and services available in open, distributed environment. In *FSEP*'s pilot system [22], agents monitors in real time the current Terminal Area Forecasts (forecasts in areas around airports) and alerts forecasters to inconsistencies between these and observations obtained from the Automatic Weather Station data.

The **CANID** system employs autonomous agents for simulating the population dynamics of coyotes using the Swarm platform (Swarm Development Group 2001). The system models territoriality and dominance of canine populations and their effects on population dynamics and supports agent interaction with variable schedules and hierarchies [65]. The model was not tied to a specific geographic area and does not account for regional differences among populations (*e.g.* litter size, pack size or territory size). Additional model development may account for this variation with changes in resources among regions.

The **NED-2** application, developed by the University of Georgia and the USDA Forest Service, deals with the simulation of forest ecosystems management plans and the evaluation of alternatives. In *NED-2* agents use growth and yield models to simulate management plans, perform goal analyzes, and generate result reports [58]. *NED-2* uses blackboard architecture and a set of semi-autonomous agents to manage the different modeling tools used.

The **PICO** project [63] adopts agent-based requirement analysis for a decision support system in the field of integrated production in agriculture. This work focuses on design issues, using Tropos methodology [29] and continuing their developments using software agents.

In **O$_3$RTAA** several software agents co-operate in a distributed agent society in order to monitor and validate measurements coming from several sensors, to assess air-quality, and to fire alarms when needed [6]. *O$_3$RTAA* relies on the agent paradigm for building intelligent software applications, while takes advantage of machine learning algorithms and data mining methodologies for extracting knowledge and customizing intelligence into agents. The system intervenes between the sensors and the experts and undertakes several tasks in order to assist humans in their evaluation. Specifically, system goals are assigned to agents that act as mediators and deliver validated information to the appropriate stakeholders.

In **AMEIM** (an *Agent-based Middleware for Environmental Information Management*) software agents undertake environmental data management tasks. The agents in *AMEIM* are capable to fuse and pre-process environmental data. *AMEIM* is a reusable platform, which realizes a generic architecture for developing agent-based systems, operating as a middleware application between environmental data pools and the final users of environmental information. Accordingly, the *AMEIM* system is fully customizable (depending on the requirements of each application) and, as mentioned before, follows an extendable architecture [7]. Reasoning capabilities can also be incorporated into *AMEIM* agents for supporting decision-support features.

DAWN (*Hybrid Agent-Based Model for Estimating Residential Water Demand*) is a simulator that integrates an agent-based social model for the consumer with conventional econometric models. It simulates the residential water demand-supply chain and thus, enables the evaluation of different scenarios for policy making. It was used to evaluate five different water-pricing policies for the period 2004-2010 in the metropolitan area of Thessaloniki [5]. Its main advantage is that it supports social interaction between consumers, through an influence diffusion mechanism, implemented via inter-agent communication (JADE and FIPA specifications).

FIRMABAR (*FIRMA* stands for *Freshwater Integrated Resource Management with Agents* and *BAR* for *Barcelona*) is an agent-based simulator, within the *FIRMA* project, aimed at simulating urban water management [42]. Such simulator provides the policy makers with an additional tool to evaluate alternative water policies in different scenarios. The simulator plays the life of a set of families (agents) on a grid that represents the territory. The global behavior of the simulation emerges as a result of the interaction of the individual agents through time (nothing in the model specifies the global-level behavior of the system). The step time in simulations is the month, and there are four central processes computed at each time step.

MANGA is a discrete event simulator (a sequential process of unrelated events) [39]. The objective of *MANGA* is to show, over a number of years (12-year period), the evolution of a group of farmer agents with a limited water resource. In *MANGA* the authors demonstrate that agent-based modeling could help negotiations by showing the consequences of water allocation rules with respect to different criteria (*e.g.* the climate of the year, the irrigated area and the level of irrigation).

MABEL (*Multi-Agent Behavioral Economic Landscape*) presents a bottom-up approach to allow the analysis of dynamic features and relations among geographic, environmental, human, and socioeconomic attributes of landowners, as well as comprehensive relational schematics of land-use change [40]. The authors adopt a distributed modeling architecture to separate the modeling of agent behaviors in Bayesian belief networks from task-specific simulation scenarios. *MABEL* has a client-server architecture, a key component that allows to simultaneously simulate land-use change over large regions in an efficient and scalable way. It separates the simulation locations from the agents' behavioral models, which simplifies the work required to parameterize these models for task specific use in the distributed modeling environment.

Control-MWS (*Agent-Based Control of a Municipal Water System*) implements a water pollution monitoring system of a simplified municipal water system (*i.e.* a single water reservoir, a single tank, a pump station with only one electrical pump, pipes and valves). It monitors the level and quality of water basically in the tanks and pumping stations, as strategic points to set up control strategies [28]. The authors use a distributed control architecture based on automation controllers with an extended firmware that supports intelligent agents. The intelligence of the system is distributed among multiple controllers by placing individual or multiple agents inside the controllers. After setting up some control strategies, simulations are done to predict the results in water quality under these control strategies.

GRENSMAAS is a project that started in the 1990s. Within the scope of this project the researchers [84] presented an agent-based model to evaluate different river management alternatives developed within the previous phases of the project. This agent-based model is coupled with an integrated river model that describes the impacts of river management, such as flood risk, nature development and costs (related to gravel extractions). Thus, the main use of the agent-based model is to investigate stakeholder environment interaction by simulating changing perspectives and behavior in response to environmental change. The agents are endowed with quantitative goal standards to evaluate their goals. The beliefs of the agents are related to their uncertainty perspectives for evaluating a river management strategy.

DSS MAS-GIS (*Decision Support System coupling Multi-Agent System and Geographical Information System*) is a framework developed to manage water in the Mediterranean islands. The MAS-GIS platform makes possible for users to better understand the current operation of the system and the evolution of the situation, while simulating different scenarios according to the selected water policies (*i.e.* best consumer water policies) and the climatic changes hypothesis [83].

PALM (*People And Landscape Model*) was used to simulate seven strategies of crop nutrient management used within a community of households (the model simulates resource flows in rural subsistence communities). *PALM* runs on a daily time step using daily weather data as driving variables. The model uses object-oriented concepts with multiple instances of various sub-models being possible. Consequently, as an example, different crop models (or even the same one) can be run simultaneously in different fields with different parameters (*e.g.* planting dates, *etc.*) for each instance. Its structure and the use of Object Oriented Programming (OOP) and agents allows a high degree of modularity, and hence flexibility ([51], [50]).

DANUBIA is a decision support system embedded in *GLOWA-Danube* project aimed at evaluating the sustainability of future water resources management alternatives, and to evaluate consequences of IPCC (Intergovernmental Panel on Climate Change) derived climate scenarios for the period from 2000 to 2100 [32]. *DANUBIA* is a coupled simulation system comprising 16 individual models [11]. To integrate the different simulation models *DANUBIA* makes use of object-oriented framework approaches. The agent-based approach, within the overall system, is used to model demography, water consumption and supply infrastructure, thus, to assess and simulate the socio-economic aspects of the water cycle (not the physical processes concerned with the water cycle). For that purpose a simulator -**DEEPACTOR**- was built providing a common conceptual and architectural basis for the modeling and implementation of the socio-economic simulation models in *GLOWA-Danube* [10].

WPMS (*Water Pollution Monitoring System*) is aimed at monitoring water quality for regulatory compliance. The water pollution monitoring system is comprised of several sites/stations in which the water quality is monitored, and when the measurements of certain parameters are exceeded, a warning is sent to the supervisor system. As the sites are geographically distributed, they are modeled in a natural way as intelligent agents that communicate with a supervisor agent who receive corresponding messages from the sites. A prototype has been designed for future implementation [60]. The system can also be used to facilitate response to contamination incidents.

Another application in coupling human and natural systems, in the area of *Land-Use Change Dynamics* (**LUCD**), is given by [53]. Land-use change dynamics were simulated for several scenarios, differentiated by the initial distribution of the different agents (*i.e.* landowner, homeowner and government types), and economic model assumptions. The goal of this work was to develop both a specific model for the study area and a general framework that captures essential features of land-use change dynamics. The used of multi-attribute key utility functions are the basis of agent rationality and decision-making.

The **SYPR** project (*Southern Yucatán Peninsular Region project*) aims at modeling and simulation of deforestation in this region. One of the main components is **HELIA** (*Human-Environment Integrated Land Assessment*). *HELIA* represents real-world households and their land-use strategies as virtual agents equipped with multi-criteria evaluation strategies and other methods (symbolic regression, and evolutionary programming). Another important component is **LU-CIM** (*Land-Use Changes In the Midwest*). The latest uses a utility-maximization approach whereby a set of household land-use preference parameters are fitted to the land-change record derived from historical aerial photography [46].

MASQUE (*Multi-Agent System for Supporting the Quest for Urban Excellence*) exploits the versatile potential of multi-agent technology for supporting the development of land-use plans [73]. It gives a detailed description of the operation of agents who are part of the system's 'knowledge' component, and then, a prototype application is developed to demonstrate how multi-agent concepts can be used to generate alternative plans. It provides functionality to make inventories of a site, *i.e.* tools to input both spatial and a-spatial data about the study area and its surroundings in order to build up project databases.

The **Thieul** simulator was developed with the help of *CORMAS* programming environment [8]. The agent-based model has been designed to formalize the interactions between the biophysics dynamics of the natural resources (*e.g.* available water, land, *etc.*) and the socio-economic factors driving the land-use dynamics around the drilling of Thieul village in the sylvo-pastoral area of Ferlo (Senegal).

SIMULAIT WATER was used to analyze urban water trading and water saving incentives among households of differing demographic types. Each agent can mimic the behavior of individual elements (*e.g.* households) in a system, as well as their interactions (*e.g.* negotiations among households). In this case, agents model individual households and their purchasing and water consumption behaviors [64].

LUDAS (*Land-Use Dynamic Simulator*) is a multi-agent system to simulate spatio-temporal dynamics of coupled human-landscape system [38]. The system is aimed at explore alternative scenarios to improve livelihoods and mitigate negative impact of land-use changes, thereby supporting the negotiation process among various stakeholders in land-use planning. Human population and the landscape environment are all self-organized interactive agents that are called upon to perform tasks in parallel (*i.e.* synchronizing actions). The framework provides a platform where many techniques already developed in spatial modeling can be integrated. For instance, the authors nested the bounded-rational decision mechanism (*e.g.* the maximization of parameterized utility functions) with the reflex mechanism (set of reflex rules) to represent the decision making mechanisms of farming households about land use.

4. Analysis and Discussion

In §3 forty-two applications using, in a more or less extent, agent-based technology in the environmental management domain, have been briefly explained. In Tables 1–5 (see pp.22–26) and 6–7 (see pp.27–28) a summary is given, respectively, together with some important characteristics used to analyze the systems reviewed. These characteristics are quoted in §3 and make reference to *agent software design* and *software development* diffusion.

From the *software design* perspective, thirty of the applications use the notion of an agent to conceptualize the system under study. These are followed by eleven applications that adopt the notion of an agent not only to conceptualize the system but to specify the software as well. Still seven of the applications reviewed use a low level notion of agents, understanding the agents as simple agent-alike "*entities*". Finally, only three of the applications adopt a more sophisticated agent-oriented software process to design the system. This latest remark suppose that, whereas from 1996 to 2004 only one of the reviewed systems had used an agent-oriented software engineering technique throughout the whole design process (*i.e. PICO*), from 2005 to 2008 two more of the studied systems have used them (*i.e. AMEIM, WPMS*). As shown in Tables 6–7 some of the systems use, in the same application, different levels of software design. If such is the case, both notions considered have been noted.

Following the same perspective (*i.e.* software design), in Tables 1–5 the agent types (or names) used to model the systems are written. They can be classified in two general categories: one group containing agents that perform specific functions (*i.e.* knowledge base, case-based reasoning, supervisory, data provider, query, broker, ontology, wrapper agents) and a second group containing agents that represent physical objects (*i.e.* pump station, watercourse agent, *etc.*), persons (*i.e.* landowner, household, farmer, taxpayer, hunting manager, *etc.*) or institutions (*i.e.* environmental lobbies, families, government agents, *etc.*). That is, in the first category there are well-known knowledge base, data mining *etc.* tasks, whereas in the second category an *agentification* of several real entities takes place. These latest agents are commonly operated with an specific model describing their behavior.

From a *software development* perspective, the applications presented are generally developed with an object-oriented language. Sixteen of them implement the system with objects. Nevertheless, twelve of them use agent-based platforms. These platforms are either generic (*e.g.* Swarm, NetLogo, *etc.*) or specific builded platforms (*e.g.* SimulaitWater, DeepActor, *etc.*). Their implementation degree, if known, is generally either partially or fully developed prototype. None of them is reported to be fully implemented as a real-time application.

When analyzing the *validation* step, twenty seven of the applications are validated. As a first-step validation, the most extended method employed to validate agent-based applications is by means of expert validation of the model they use. As a second step, most of the applications permit to do some simulations and to compare the simulated results against historical and/or observed data (when available). Few of them use other, more sophisticated, techniques (*i.e.* sensitivity analysis, extreme tests or cases).

5. Conclusions

The *state of the art* in agent-based approaches applied to environmental issues shows the utility of agents as solvers of environmental problems. The applications and agents used are heterogeneous in nature: although most of them refer to natural resources management (*i.e.* from water sources, air or soil), other environmental issues are also faced using agents. Their coupled work permit to go beyond their individual capabilities or knowledge. All these applications have some of the general characteristics of multi-agent systems reported in [79]:

- Each agent has incomplete information or capabilities for solving the problem. Thus, the importance of MASs is concerned with the behavior of a collection of agents designed at solving a given problem together.
- There is no global system control.
- Data is decentralized, and
- Computation is asynchronous.

The design of the systems studied is mainly done using agent-based concepts whereas for their implementation the use of object-oriented technologies prevails. The systems are partially validated: in most of the cases the model used to describe the agents is validated through expert knowledge, whereas the overall system performance validation is a further step that requires more work and research to be done.

As concluded in [4], agent-based technology is not homogeneously adopted in environmental software developments. However, an increase in the use of agent platforms to develop the systems is observed. Even though the fuzzy classification of the systems into the three groups described in §3 (*i.e.* EDMS, EDSS, ESS), no interrelation between the type of agent-based environmental system and the technology used can be observed.

Design and implementation of MASs aimed at solving environmental problems require research in order to tackle with many challenges. Some of the most important and tricky ones were listed in [35], and are still important questions by researchers in the field of multi-agent system applications. Answers to these questions are naturally interrelated. Some answers are found within the reviewed systems presented in this *state of the art* and some others are offered in this book for some specific environmental problems.

Acronym	Main tasks and objectives	Application Field	Related technologies	Agents (names or types)
DAI-DEPUR [74][75]	Simulation and control of the physical, chemical, microbiological aspects of the activated sludge processes	Wastewater treatment plants	LISP, G2, GAR, LINNEO+	Knowledge base agents, case-base reasoning agents, supervisory agents
EDS-DAI [44]	Project evaluation and assessment with respect to alternative locations that comply with legal regulations, development plans and satisfy custom requirements	Environmental project evaluation	Distributed Belief Revision, ARCHON	Decision Support Agents (or evaluation agents), Data Provider Agents (stored in GIS), User Interface Agent
SAEM [76]	Monitoring the pollutant cloud emitted by a power plant chimney	Atmospheric pollution	Robotic agents	Helicopter agents
ESAT-WMR [23][24]	Modeling and analysis of elective strategies for urban water supply pipe network rehabilitation	Water supply networks	KQML, Object-oriented programming	Interface agent, Heuristics agent, Information agent, Data mining agent, Database agent, Constraint Agent, Predictor Agent, HotSpot Agent
IDS-DAP [49][48]	market penetration of agricultural products investigation, using multicriteria analysis	Differentiated agricultural products marketing	UML, Visual Basic, TCP-IP, Java	Data analysis agent, Brand Choice agent, Market expert agent
FIRMA & Thames [54][12]	Agent-based modeling for the integration of natural, hydrologic, social and economic aspects of freshwater management.	Water resource management	SDML	Policy agent, Citizens
SHADOC [9]	Farmer behavior and water allocation simulation	Water catchment management	UML, SmallTalk, Object-oriented programming, Petri Nets, CORMAS, Visual Works®	PumpStation, Reach, Water-course, Farmer agents
EDEN-IW & InfoSleuth [57][66][25]	Data integration and homogeneous access provision services	Water resources data	JADE, FIPA-ACL, SQL, RDF, OKBC	DB resource agent, query decomposition agent, ontology agent, broker agent

TABLE 1. Summary of the reviewed systems.

Acronym	Main tasks and objectives	Application Field	Related technologies	Agents (names or types)
WaWAT (WaWo) [18] [19]	A multi-agent cooperation infrastructure for supervision and decision-making in wastewater treatment plants	Wastewater treatment plants	Ontolingua KSL	Dynamic *entities* (monitoring, modeling, actuator, predictive agents, *etc.*)
BUSTER [56]	Data integration and filtering, querying services	Geographical information sources	OIL, FIPA-OS	Wrapper, mediator, mapper
Adour [81]	Stakeholder negotiation over water use	Water management	BDI	Farmers, environmental lobbies, water manager, taxpayer
MAGIC & DIAMOND [37] [2]	Fault detection in industrial process	Water treatment process and Water steam cycle a power plant	XML, CORBA, FIPA-ACL	Diagnostic agents, data acquisition agents, knowledge acquisition agent, wrapper agents, monitoring agent
NZDIS [21] [68]	Integrated querying services in an open, distributed environment of heterogeneous databases	Environmental data	FIPA-ACL, UML, OQL, RDF	Ontology agent, resource agent, query processing agents, broker agent
D-NEMO [36]	Air pollution incident forecasting	Atmospheric pollution	LALO, KQML	Station agents, model agents
RAID [52]	pollution monitoring and control in indoor environments	Indoor air quality	UML, Kaleidoscope	*entities*: manager, sensors, *etc.*
AdEcAA [70]	simulation of aquatic food webs and plankton species interactions	Food chain	Echo	Phytoplankton species, zooplankton species
CATCHSCAPE [13]	Simulation of the whole catchment features as well as farmer's individual decisions	Water catchment management	UML, SmallTalk, Object-oriented programming, CORMAS	Crop, Farmer, Canal, Weir, Canal Manager, River
SINUSE [26]	physical and socio-economic interactions modelling for simulating demand management of a water table	Integrated management of a water table	UML, SmallTalk, Object-oriented programming	Plot, Water table, Farmer
STAU-Wien [41]	Simulation or rural development patterns in the Vienna Region	Rural development	UML, ArcInfo, Cellular automata, Object-oriented programming	Enterprises, households

TABLE 2. (continued) Summary of the reviewed systems.

Acronym	Main tasks and objectives	Application Field	Related technologies	Agents (names or types)
GEMACE [47]	Simulation of interactions between duch population, farming decisions and leasing of hunting rights	Environmental Object-oriented planning	UML, SmallTalk, CORMAS	Hunting manager agents, farmer agents
FSEP [22]	Surveillance, forecasting and alert of weather conditions	Meteorology	JACK, RDF-S, DAML+OIL	Wrapper agents, interface agents
CANID [65]	Agent-based simulation of territoriality and cominance of canid populations	Biodiversity-population dynamics	Swarm	coyote, pack
NED-2 [58]	Forest ecosystem management simulation and goal-driven decision support	Forest management	C++, HTML	Interface agent, Simulation agent, Goal analysis, planning agent
PICO [63]	Design system requirements, analysis of organizational complexity, dealing with all the dependencies between the domain stakeholders, and study of natural plant protection techniques.	Integrated production in agriculture	Tropos, WEKA	GIS agents, Disease Behavior Learner, wrapper agents
O$_3$RTAA [6]	A MAS for monitoring and assessing air-quality attributes, firing alarms to appropiate recipients when needed.	Urban Pollution Control (air)	JADE, FIPA-ACL, Protégé 2000 (exported in RDFS), WEKA, JESS, PMML	Diagnosis Agent, Database Agent, Distribution Agent, Alarm Agent
AMEIM [7]	A MAS able to capture and validate environmental data from several external sources.	Environmental data monitoring and management	GAIA, AORML, JADE, Protégé 2000, FIPA-ACL	Contribution Agents, Data Management Agents, Distribution Agents, Graphical User Interface Agent
DAWN [5]	Simulation of residential water demand and how water pricing policies affect demand	Water demand management	AORML, 2D grids, JADE	Simulator, meteo, consumer, supplier agents
FIRMABAR [42]	Integrated freshwater assessment in a geographic area by means of water supply/demand simulations (in different scenarios)	Urban water management	SDML, Swarm libraries (Java), OOP	Families, companies, municipalities, government agents

TABLE 3. (continued) Summary of the reviewed systems.

Acronym	Main tasks and objectives	Application Field	Related technologies	Agents (names or types)
MANGA [39]	Simulation of decision-making process and of the impact of water allocation on farmer's collective behavior	Rural development, water resources management	UML	Farmers, water suppliers, crops, climate, information supplier agents
MABEL [40]	Simulation of land-use changes over time and space	Land use	BDI architecture, Swarm, VisualStudio.NET C/C++, BBN (Bayesan Belief Network)	Policy maker, landowners (farmer agent, ubrban residential agent, forestry agent, household agent, etc.)
Control-MWS [28]	Water pollution monitoring system (water quality, energy costs and demand) of a simplified municipal water system	Urban water data management	Simulink tool	Pumping station, tank agents
GRENSMAAS project [84]	Simulation of stakeholder support in under different policy strategies (nature development, gravel extraction, flood reduction)	Water catchment management	BDI (approach)	Policy makers, citizens, farmers, nature organizations, gravel extractors agents
MAS-GIS DSS [83]	Decision support system framework for water management in the Mediterranean islands coupling a Multi-Agents System with a Geographic Information System	Water management	CORMAS, ARCGIS, ODBC	Drillings, tanks, water companies, consumers (hotels and homes), and a water police agents
PALM [50] [51]	Simulation of management strategies in a community of households in Nepal (linking decision-making to underlying biological processes in soil nutrient dynamics)	Rural development	UML, OOP	Household, landscape, livestock agents
DANUBIA (DEEP-ACTOR) [11] [10]	Simulation of scenarios and strategies for the future of water in the upper Danube Basin (an integrative decision support system)	Water resources management (water supply and groundwater) under conditions of global change	UML, object-oriented programming (OOP)	Farmer agents (maize, Meat-Breed, etc.), WaterSupply-Company and Household agents

TABLE 4. (continued) Summary of the reviewed systems.

Acronym	Main tasks and objectives	Application Field	Related technologies	Agents (names or types)
WPMS [60]	Water pollution monitoring for regulatory compliance (early stage of research; analysis phase)	Urban water management	FIPA-ACL, UML, agents	Monitor, supervisor and control agents
Thieul [8]	Simulation an agro-sylvopastoral context	Integrated natural resource management	UML, CORMAS	Farmers, herders and farmer-herders agents
LUCD [53]	Determining conditions of the interactions between human decisions and natural systems that lead to long-term sustainability of forest ecosystems	Land-use management, forest management	multi-attribute utility functions	Landowner, developer, home-owner, government agents
SYPR project (HELIA and LUCIM) [46]	Modeling land change and economic decision-making in the United States (LUCIM) and Mexico (HELIA)	Land-use management, forest management	evolutionary programming, multi-criteria evaluation, symbolic regression	Households agents (agriculturalists types)
MASQUE [73]	Multi-agent planning support system that supports decisions related to complex, uncertain and subjective urban planning problems.	Land use (urban)	BDI, UML, Borland JBuilderTM	Facilitation, interface, tool and domain (refer to land-use) agents
SIMULAIT WATER [64]	Simulation and analysis of various pricing and trading policies	Urban water management (supply and trading)	Scripting language	Household agents (low, medium and high)
LUDAS [38]	Spatio-temporal simulation of a coupled human landscape system.	Land-use and rural development	NetLogo 3.0	Household, landscape, agricultural agents

TABLE 5. (continued) Summary of the reviewed systems.

Acronym	SW Design	SW Dev.	Implementation Degree	Validation
DAI-DEPUR	2,3	D	Partial. The rule-based component and the case-based component were implemented, but not interconnected. It was continued in the WaWAT (WaWo) system. Also, a real-world application was delivered in the ATL-EDAR system [75].	Is incrementally being done at several points during its development. Whole system validation at three levels (1) simulation of the plant in real time, (2) building-up and testing on a pilot scale plant, and (3) validation on a real plant
EDS-DAI	2	C	The system prototype is under development	Two stages of evaluation : (1) Submission to the relevant group of public (and private) agencies, (2) Incorporation of consulted agencies' opinions
SAEM	2	D	Unkwnon	The use of simulation gives the chance of testing this kind of behaviours without building the real agents
ESAT-WMR	3	D	Partial	No
IDS-DAP	2	A	Unknown	No
FIRMA & Thames	1, 2	A	Full	Validation of model struct. and simulation results with stakeholders (focus groups) (comp. validation)
SHADOC	2	A	Full	Expert validation
EDEN-IW & InfoSleuth	2, 3	C	Partial (EDEN-IW DEMO available)	No
WaWAT (WaWo)	2,3	D	A prototype	Through some case study
BUSTER	2	B	A first prototype	No
Adour	3		Future Implementation in a case study (Adour Basin)	No
MAGIC & DIA-MOND	2, 3	A, B	Core toolkit developed	Evaluation examples. Comparison values with simulated offline and online ones
NZDIS	2	B	Full	Unknown
D-NEMO	3	C	Full	Experimental multiagent prototype under simulated real time conditions
RAID	1	A	Unknown	Unknown
AdEcAA	2	A	An example of individual-based adaptative agents simulation system is implemented on the Echo framework	Through a multivariate time-series database for nine lakes different in climate, eutrophication and morphology

TABLE 6. Deep analysis of the reviewed systems.

Acronym	SW Design	SW Dev.	Implementation Degree	Validation
CATCH-SCAPE	2	A	Some prototypes	Comparison of the average simulated yields with those provided by local Thai Agencies
SINUSE	2	A	Full	Two step validation: 1) Extreme tests, 2) Partial sensitivity analysis
STAU-Wien	2	A	Full	No
GEMACE	2	A	Some prototypes	Expert validation
FSEP	2	C	A prototype	Through comparison between observed and forecasted data
CANID	2	C	Unknown	Comp. with other models; sensitivity analysis and calibration methods
NED-2	2	A, D	A prototype	Planned
PICO	4		No	No
O₃RTAA	2, 3	C	Full	In a single meteorological station. Extended validation planned.
AMEIM	4	C	Full (AMEIM ver.1.0)	Unknown
DAWN	2	C	Full	Metropolitan Area of Thessaloniki (under 5 scenarios). Expert validation
FIRMA-BAR	2	C	Full	Barcelona and Valladolid (under several scenarios). Expert validation
MANGA	2	A	Full	Qualitative
MABEL	3	C	Full	Against historical data
Control-MWS	1	A	Full	In a municipal wastewater system
GRENS-MAAS	3	A	Partia	Comparison with historical data
MAS-GIS DSS	1	C	A prototype	Expert validation
PALM	2	A	Partial	Two step validation: 1) Comp. with historical data, 2) Expert validation
DANUBIA (DEEP-ACTOR)	2	D	Full (not yet avail. for the interested end users, *i.e.* governm. institutions)	Two step validation: 1) Comparison with observed values; 2) Expert validation
WPMS	4		No	No
Thieul	2	C	Full	Expert validation
LUCD	1, 2		Full (optimization of utility functions)	Comparison with real data
SYPR (HE-LIA and LUCIM)	1, 2		Full (optim. of utility funct. and use of multi-criteria, symb. regression and evol. progr.)	Comparison of experimental data with expert knowledge
MASQUE	3	A	A prototype	Planned
SIMULAIT WATER	1, 2	D	A prototype	No
LUDAS	2	C	Full	Model validation in progress

TABLE 7. (continued) Deep analysis of the systems reviewed.

References

1. L. Adelman, *Evaluating decision support and expert systems*, Wiley-Interscience New York, NY, USA, 1992.

2. M. Albert, T. Laengle, H. Woern, M. Capobianco, and A. Brighenti, *Multi-agent systems for industrial diagnostics*, Proceedings of 5th IFAC Symposium on Fault Detection, Supervision and Safety of Technical Processes. Washington DC, USA, June 9-11 **3** (2003).

3. R. Annicchiarico, U. Cortés, and C. Urdiales, *Agent Technology and e-Health*, Whitestein Series in Software Agent Technologies and Autonomic Computing, Birkhäuser Verlag, 2008.

4. I.N. Athanasiadis, *A review of agent-based systems applied in environmental informatics*, In A. Zerger and R.M. Argent, editors, *MODSIM 2005 Int'll Congress on Modelling and Simulation*, Modelling and Simulation Society of Australia and New Zealand, Melbourne, Australia, December 2005, pp. 1574–1580.

5. I.N. Athanasiadis, A.K. Mentes, P.A. Mitkas, and Y.A. Mylopoulos, *A Hybrid Agent-Based Model for Estimating Residential Water Demand*, SIMULATION **81** (2005), no. 3, 175.

6. I.N. Athanasiadis and P.A. Mitkas, *An agent-based intelligent environmental monitoring system*, Management of Environmental Quality **15** (2004), no. 3, 238–249.

7. I.N. Athanasiadis, A. Solsbach, J. Marx Gómez, and P. Mitkas, *An Agent-based Middleware for Environmental Information Management*, 2nd Conf. on Information Technologies in Environmental Engineering (ITEE'2005) (2005), 253–267.

8. A. Bah, I. Touré, C. Le Page, A. Ickowicz, and A.T. Diop, *An agent-based model to understand the multiple uses of land and resources around drillings in Sahel*, Mathematical and Computer Modelling **44** (2006), no. 5-6, 513–534.

9. O. Barreteau and F. Bousquet, *SHADOC: a multi-agent model to tackle viability of irrigated systems*, Annals of Operations Research **94** (2000), no. 1, 139–162.

10. R. Barthel, S. Janisch, N. Schwarz, A. Trifkovic, D. Nickel, C. Schulz, and W. Mauser, *An integrated modelling framework for simulating regional-scale actor responses to global change in the water domain*, Environmental Modelling and Software (2008).

11. R. Barthel, V. Rojanschi, J. Wolf, and J. Braun, *Large-scale water resources management within the framework of GLOWA-Danube. Part A: The groundwater model*, Physics and Chemistry of the Earth **30** (2005), no. 6-7, 372–382.

12. O. Barthelemy, S. Moss, T. Downing, and J. Rouchier, *Policy modelling with ABSS: The case of water demand management*, CPM Report (2001).

13. N. Becu, P. Perez, A. Walker, O. Barreteau, and C.L. Page, *Agent based simulation of a small catchment water management in northern Thailand. Description of the CATCHSCAPE model*, Ecological Modelling **170** (2003), no. 2-3, 319–331.

14. R.J. Boland Jr, A.K. Maheshwari, D. Te'eni, D.G. Schwartz, and R.V. Tenkasi, *Sharing perspectives in distributed decision making*, Proceedings of the 1992 ACM conference on Computer-supported cooperative work, ACM New York, NY, USA, 1992, pp. 306–313.

15. E. Bonabeau, F. Henaux, S. Guerin, D. Snyers, P. Kuntz, and G. Theraulaz, *Routing in Telecommunications Networks with Ant-Like Agents*, Lecture Notes in Computer Science (1998), 60–71.

16. B. Brehmer, *Distributed decision making: some notes on the literature*, Distributed Decision Making: Cognitive Models for Cooperative Work, John Wiley & Sons, London (1991), 3–14.

17. M. Caridi and S. Cavalieri, *Multi-agent systems in production planning and control: an overview*, Production Planning and Control **15** (2004), no. 2, 106–118.

18. L. Ceccaroni, *What If a Wastewater Treatment Plant Were a Town of Agents*, Proceedings of the workshop Autonomous Agents 2001-W03: Ontologies in Agent Systems, Montréal, Canadà (2001).

19. L. Ceccaroni, U. Cortés, and M. Sànchez-Marrè, *WaWO-An ontology embedded into an environmental decision-support system for wastewater treatment plant management*, Proceedings of ECAI2000-W09: Applications of ontologies and problem-solving methods (2000), 2–1.

20. A. Chávez and P. Maes, *Kasbah: An agent marketplace for buying and selling goods*, First International Conference on the Practical Application of Intelligent Agents and Multi-Agent Technology (PAAM96) (1996), 75–90.

21. S. Cranefield and M. Purvis, *Integrating environmental information: incorporating metadata in a distributed information system's architecture*, Advances in Environmental Research **5** (2001), no. 4, 319–325.

22. S. Dance, M. Gorman, L. Padgham, and M. Winikoff, *An evolving multi agent system for meteorological alerts*, Proceedings of the second international joint conference on Autonomous agents and multiagent systems, AAMAS-03 (2003), 966–967.

23. D. Davis and B. Sharp, *Application of expert system and agent technology to water mains rehabilitation decision making*, New Review of Applied Expert Systems **5** (1999), 5–18.

24. D.N. Davis, *Agent-based decision-support framework for water supply infrastructure rehabilitation and development*, Computers, Environment and Urban Systems **24** (2000), no. 3, 173–190.

25. B. Felluga, T. Gauthier, A. Genesh, P. Haastrup, C. Neophytou, S. Poslad, D. Preux, P. Plini, I. Santouridis, M. Stjernholm, et al., *Environmental data exchange for inland waters using independed software agents, Report 20549 EN*, Institute for Environment and Sustainability, European Joint Research Centre, Ispra, Italy (2003).

26. S. Feuillette, F. Bousquet, and P. Le Goulven, *SINUSE: a multi-agent model to negotiate water demand management on a free access water table*, Environmental Modelling and Software **18** (2003), no. 5, 413–427.

27. J. Fox and S. Das, *Safe and sound: artificial intelligence in hazardous applications*, AAAI Press/The MIT Press. Cambridge, MA, USA, 2000.

28. L. Giannetti, F.P. Maturana, and F.M. Discenzo, *Agent-based control of a municipal water system*, Lecture Notes in Computer Science (2005), 500–510.

29. F. Giunchiglia, J. Mylopoulos, and A. Perini, *The Tropos Software Development Methodology: Processes, Models and Diagrams*, In Proceedings of the first international joint conference on Autonomous agents and multiagent systems: part 1, **15** (2002), no. 19, 35–36.

30. I. Graham and P.L. Jones, *Expert systems: knowledge, uncertainty and decision*, Chapman and Hall. New York, 1988.

31. G. Guariso and H. Werthner, *Environmental decision support systems*, Ellis Horwood-Wiley, New York, 1989.

32. J.T. Houghton, Y. Ding, D.J. Griggs, M. Noguer, P.J. van der Linden, X. Dai, K. Maskell, and C.A. Johnson, *Climate change 2001: the scientific basis. Intergovernmental Panel on Climate Change*, Cambridge University Press: Cambridge. The Independent, Apr **14** (2001), 2006.

33. J. Huang, N.R. Jennings, and J. Fox, *Agent-based approach to health care management*, Applied Artificial Intelligence **9** (1995), no. 4, 401–420.

34. N.R. Jennings, P. Faratin, M.J. Johnson, T.J. Norman, P. O'Brien, and M.E. Wiegand, *Agent-Based Business Process Management*, IJCIS **5** (1996), no. 2&3, 105–130.

35. N.R. Jennings, K. Sycara, and M. Wooldridge, *A Roadmap of Agent Research and Development*, Autonomous Agents and Multi-Agent Systems **1** (1998), no. 1, 7–38.

36. E. Kalapanidas and N. Avouris, *Air Quality Management Using a Multi-Agent System*, Computer-Aided Civil and Infrastructure Engineering **17** (2002), no. 2, 119–130.

37. B. Köppen-Seliger, S.X. Ding, and P.M. Frank, *European research projects on multi-agents-based fault diagnosis and intelligent fault tolerant control*, Plenary Lecture IAR Annual Meeting, Strasbourg (2001).

38. Q.B. Le, S.J. Park, P.L.G. Vlek, and A.B. Cremers, *Land-Use Dynamic Simulator (LUDAS): A multi-agent system model for simulating spatio-temporal dynamics of coupled human–landscape system. I. Structure and theoretical specification*, Ecological Informatics **3** (2008), 135–153.

39. M. Le Bars, J.M. Attonaty, S. Pinson, and N. Ferrand, *An Agent-Based Simulation Testing the Impact of Water Allocation on Farmers' Collective Behaviors*, SIMULATION **81** (2005), no. 3, 223.

40. Z. Lei, B.C. Pijanowski, K.T. Alexandridis, and J. Olson, *Distributed Modeling Architecture of a Multi-Agent-Based Behavioral Economic Landscape (MABEL) Model*, SIMULATION **81** (2005), no. 7, 503.

41. W. Loibl and T. Toetzer, *Modeling growth and densification processes in suburban regions – simulation of landscape transition with spatial agents*, Environmental Modelling and Software **18** (2003), no. 6, 553–563.

42. A. López-Paredes, D. Saurí, and J.M. Galan *Urban Water Management with Artificial Societies of Agents: The FIRMABAR Simulator*, SIMULATION **81** (2005), no. 3, 189.

43. M. Luck, P. McBurney, O. Shehory, and S. Wilmott, *Agent Technology Roadmap: A Roadmap for Agent Based Computing*, AgentLink Community (2005).

44. B. Malheiro and E. Oliveira, *Environmental decision support: A distributed artificial intelligence approach*, Proc. of the International Symposium and Workshop: Environment and Interaction (1996).

45. ———, *Environmental decision support: a multi-agent approach*, International Conference on Autonomous Agents: Proceedings of the first international conference on Autonomous Agents **5** (1997), no. 08, 540–541.

46. S.M. Manson and T. Evans, *Agent-based modeling of deforestation in southern Yucatan, Mexico, and reforestation in the Midwest United States.*, Proc Natl Acad Sci USA **104** (2007), no. 52, 20678–83.

47. R. Mathevet, F. Bousquet, C. Le Page, and M. Antona, *Agent-based simulations of interactions between duck population, farming decisions and leasing of hunting rights in the Camargue (Southern France)*, Ecological Modelling **165** (2003), no. 2-3, 107–126.

48. N. Matsatsinis, P. Moraïtis, V. Psomatakis, and N. Spanoudakis, *An Agent-based System for Products Penetration Strategy Selection*, Applied Artificial Intelligence **17** (2003), no. 10, 901–925.

49. N.F. Matsatsinis, P.N. Moraïtis, V.M. Psomatakis, and N.I. Spanoudakis, *Towards an intelligent decision support system for differentiated agricultural products*, Proc. of the 5th International Conference of the Decision Sciences Institute (1999).

50. R. Matthews, *The People and Landscape Model (PALM): Towards full integration of human decision-making and biophysical simulation models*, Ecological Modelling **194** (2006), no. 4, 329–343.

51. R.B. Matthews and C. Pilbeam, *Modelling the long-term productivity and soil fertility of maize/millet cropping systems in the mid-hills of Nepal*, Agriculture, Ecosystems and Environment **111** (2005), no. 1-4, 119–139.

52. D. Micucci, *Exploiting the Kaleidoscope architecture in an industrial environmental monitoring system with heterogeneous devices and a knowledge-based supervisor*, Proceedings of the 14th International Conference on Software Engineering and Knowledge Engineering (2002), 685–688.

53. M. Monticino, M. Acevedo, B. Callicott, T. Cogdill, and C. Lindquist, *Coupled human and natural systems: A multi-agent-based approach*, Environmental Modelling and Software **22** (2007), no. 5, 656–663.

54. S. Moss, T. Downing, and J. Rouchier, *Demonstrating the Role of Stakeholder Participation: An Agent Based Social Simulation Model of Water Demand Policy and Response*, CPM Report No. 00-76, Centre for Policy Modelling, The Business School, Manchester Metropolitan University, Manchester, UK (2000).

55. J.L. Nealon and A. Moreno, *Agent-Based Applications in Health Care*, Applications of Software Agent Technology in the Health Care Domain (A. Moreno and J.L. Nealon, eds.), Whitestein Series in Software Agent Technologies, Birkhäuser Verlag, 2003, pp. 3–18.

56. H. Neumann, G. Schuster, H. Stuckenschmidt, U. Visser, T. Vögele, and H. Wache, *Intelligent brokering of environmental information with the BUSTER system*, International Symposium Informatics for Environmental Protection **30** (2001), 505–512.

57. M. Nodine, J. Fowler, T. Ksiezyk, B. Perry, M. Taylor, and A. Unruh, *Active Information Gathering in InfoSleuth*, International Journal of Cooperative Information Systems **9** (2000), no. 1/2, 3–28.

58. D. Nute, W.D. Potter, F. Maier, J. Wang, M. Twery, H.M. Rauscher, P. Knopp, S. Thomasma, M. Dass, H. Uchiyama, et al., *NED-2: an agent-based decision support system for forest ecosystem management*, Environmental Modelling and Software **19** (2004), no. 9, 831–843.

59. H.S. Nwana, *Software agents: An overview*, Knowledge Engineering Review **11** (1996), no. 3, 205–244.

60. M. Oprea and C. Nichita, *Applying Agent Technology in Water Pollution Monitoring Systems*, Proceedings of the Eight International Symposium on Symbolic and Numeric Algorithms for Scientific Computing (SYNASC'06). (2006), 233–238.

61. H.v.D. Parunak, *Agents in overalls: Experiences and issues in the development and deployment of industrial agent-based systems*, International Journal of Cooperative Information Systems **9** (2000), no. 3, 209–227.

62. H.v.D. Parunak, A. Ward, M. Fleischer, and J. Sauter, *A Marketplace of Design Agents for Distributed Concurrent Set-Based Design*, Advances in Concurrent Engineering, CE 97 (1997).

63. A. Perini and A. Susi, *Developing a decision support system for integrated production in agriculture*, Environmental Modelling and Software **19** (2004), no. 9, 821–829.

64. D. Perugini, M. Perugini, and M. Young, *Water saving incentives: An agent-based simulation approach to urban water trading*, In Simulation Conference: Simulation - Maximising Organisational Benefits (SimTecT 2008), Melbourne, Australia, May 12 - 15, 2008.

65. W.C. Pitt, P.W. Box, and F.F. Knowlton, *An individual-based model of canid populations: modelling territoriality and social structure*, Ecological Modelling **166** (2003), no. 1-2, 109–121.

66. G. Pitts and J. Fowler, *InfoSleuth: An emerging technology for sharing distributed environmental information*, In Information Systems and the Environment, 159–172.

67. M. Poch, J. Comas, I. Rodríguez-Roda, M. Sànchez-Marrè, and U. Cortés, *Designing and building real environmental decision support systems*, Environmental Modelling and Software **19** (2004), no. 9, 857–873.

68. M. Purvis, S. Cranefield, R. Ward, M. Nowostawski, D. Carter, and G. Bush, *A multi-agent system for the integration of distributed environmental information*, Environmental Modelling and Software **18** (2003), no. 6, 565–572.

69. A.S. Rao and M.P. Georgeff, *BDI Agents: From Theory to Practice*, Proceedings of the First International Conference on Multi-Agent Systems (ICMAS-95) (1995), 312–319.

70. F. Recknagel, *Simulation of aquatic food web and species interactions by adaptive agents embodied with evolutionary computation: a conceptual framework*, Ecological Modelling **170** (2003), no. 2-3, 291–302.

71. A.E. Rizzoli and W.J. Young, *Delivering environmental decision support systems: software tools and techniques*, Environmental Modelling and Software **12** (1997), no. 2-3, 237–249.

72. S.J. Russell and P. Norvig, *Artificial intelligence: A modern approach*, 2nd ed., Pearson Education, Inc., 2003.

73. J.M. Saarloos, T.A. Arenzte, A.W.J. Borgers, and H.J.P. Timmermans, *A multi-agent paradigm as structuring principle for planning support systems*, Computers, Environment and Urban Systems **32** (2008), no. 1, 29–40.

74. M. Sànchez-Marrè, U. Cortés, J. Lafuente, I. Rodríguez-Roda, and M. Poch, *DAI-DEPUR: an integrated and distributed architecture for wastewater treatment plants supervision*, Artificial Intelligence in Engineering **10** (1996), no. 3, 275–285.

75. M. Sànchez-Marrè, M. Martínez, I. Rodríguez-Roda, J. Alemany, and U. Cortés, *Using CBR to improve intelligent supervision and management of wastewater treatment plants: the atl-EDAR system*, In proceedings of 7th European Conference on Case-Based Reasoning (ECCBR'2004) (Industrial day), Madrid (2004), 79–91.

76. J.C. Seco, C. Pinto-Ferreira, and L. Correia, *A Society of Agents in Environmental Monitoring*, From Animals to Animats 5: Proceedings of the Fifth International Conference on Simulation of Adaptive Behavior (1998).

77. R.S. Sojda, *Artificial Intelligence Based Decision Support for Trumpeter Swan Management*, Ph.D. thesis, Colorado State University, 2002.

78. R.H. Sprague Jr. and E.D. Carlson, *Building Effective Decision Support Systems*, Prentice Hall Professional Technical Reference, 1982.

79. K.P. Sycara, *Multiagent Systems*, AI Magazine **19** (1998), no. 2, 79–92.

80. K. Takahashi, Y. Nishibe, I. Morihara, and F.I. Hattori, *Intelligent pages: collecting shop and service information with software agents*, Applied Artificial Intelligence **11** (1997), no. 6, 489–499.

81. S. Thoyer, S. Morardet, P. Rio, L. Simon, R. Goodhue, and G. Rausser, *A bargaining model to simulate negotiations between water users*, Journal of Artificial Societies and Social Simulation **4** (2001), no. 2.

82. A. Tolk, *An Agent-Based Decision Support System Architecture for the Military Domain*, In: Phillips-Wren, G.E. and Jain, L.C., Eds., Agent-Mediated Environments in Intelligent Decision Support Systems (2005).

83. D. Urbani and M. Delhom, *Water Management Policy Selection Using a Decision Support System Based on a Multi-agent System*, Lecture Notes in Computer Science **3673** (2005), 466–469.

84. P. Valkering, J. Rotmans, J. Krywkow, and A. van der Veen, *Simulating Stakeholder Support in a Policy Process: An Application to River Management*, SIMULATION **81** (2005), no. 10, 701.

85. R. Weihmayer and H. Velthuijsen, *Intelligent agents in telecommunications*, Springer-Verlag, 1998.

86. M. Wooldridge, *Introduction to Multiagent Systems*, John Wiley & Sons, Inc. New York, NY, USA, 2001.

87. M. Wooldridge, N.R. Jennings, and D. Kinny, *The Gaia Methodology for Agent-Oriented Analysis and Design*, Autonomous Agents and Multi-Agent Systems **3** (2000), no. 3, 285–312.

88. F. Zambonelli, N.R. Jennings, and M. Wooldridge, *Developing multiagent systems: The Gaia methodology*, ACM Transactions on Software Engineering and Methodology (TOSEM) **12** (2003), no. 3, 317–370.

Montse Aulinas
Laboratory of Chemical and Environmental Engineering
Scientific and Technological Park (University of Girona)
Edifici Jaume Casademont
Pic de Peguera, 15
Girona, 17071
Catalonia
e-mail: montseaulinas@gmail.com

Clàudia Turon
Consorci per a la Defensa de la Conca del Riu Besòs
Avinguda Sant Julià, 241
Granollers 08403
Catalonia
e-mail: cturon@besos.cat

Miquel Sànchez-Marrè
Knowledge Engineering and Machine Learning Group (KEMLG)
Software Department
Technical University of Catalonia
Jordi Girona 1-3
Barcelona 08034
Catalonia
e-mail: miquel@lsi.upc.edu

Whitestein Series in Software Agent Technologies and Autonomic Computing, 37–60

Deliberation about the Safety of Industrial Wastewater Discharges into Wastewater Treatment Plants

Pancho Tolchinsky, Montse Aulinas, Ulises Cortés and Manel Poch

Abstract. The daily operation of Wastewater Treatment Plants (WWTPs) in industrialized areas is of particular concern because of the severe problems that can occur in the WWTP caused by the incoming inflow, which in turn may cause an ecological imbalance in the fluvial ecosystem. In order to minimize the environmental impact caused by the industrial wastewater discharges, guidelines and regulations exists. However, due to the complexity of the domain, there are still no *golden standards* by which to decide whether a WWTP can cope with wastewater discharges, and so strict adherence to regulations may not always be convenient. Special circumstances may motivate to accept discharges that are above established thresholds or to reject discharges that comply with guidelines. Nonetheless, because of the criticality of the actions to be taken, such decisions require to be well justified. Hence, in this work it is proposed the use of the argumentation-based model *Pro-CLAIM* to provide a more flexible decision making process, in which expertise can deliberate whether an industrial wastewater can safely be discharged into a WWTP, and thus adapt each decision to the particular circumstance. To ensure a safe decision, agents' given arguments for or against the industrial spill are evaluated accounting for the domain guidelines and regulations, for similar past cases and for confidence in the expertise's assessments.

Keywords. Multi-Agent Systems, Argumentation, Integrated Wastewater Management.

1. Introduction

The daily operation of Wastewater Treatment Plants (WWTPs) in industrialized areas is of particular concern because of the severe problems that can occur in the WWTP caused by the incoming inflow, particularly when produced by these industries. Because problems in the WWTP can in turn cause sever ecological imbalances in the fluvial ecosystem, there is a need to regulate the allowed industrial wastewater discharges into WWTP. However, due to the complexity of the domain it is difficult to define any *golden standard* by which to decide whether a WWTP can cope with a wastewater discharge. Thus, while regulations and guidelines exists governing such decisions, their strict adherence may not always be convenient. Special circumstances may motivate either to accept discharges that are above established thresholds or to reject discharges that comply with guidelines. Nonetheless, because of the criticality of the actions to be taken, such decisions require to be well justified. Hence, in this work it is proposed the use of the argumentation-based model *ProCLAIM* to formalize a more flexible decision making process (see §4), in which expertise can deliberate whether an industrial wastewater can safely be discharged into a WWTP. The deliberation is defined as an argumentative process for eliciting the relevant factors from the decision makers (experts in the domain). This process results in a network of interacting arguments in favor or against the industrial discharge. These arguments are then evaluated accounting for the domain's guidelines and thresholds, the available empirical evidence and the decision makers' opinion. The result of such an evaluation is a justification why the proposed industrial spill may or may not environmentally safe. Moreover, the conceptual model presents a level of *computationality* that makes the implementation feasible.

In §2, we describe the context in which the decision making process takes place, motivating a need for an alternative representation of the problem. This alternative approach enables deliberation about the safety of an industrial discharge. §3 introduces the basic concepts of argumentation and §4 presents the overall model to discuss about the safety of a discharge beyond the existing standards. §5 focuses on the task of guiding the agent's argument submission, that is, the deliberation aspect of the model. In §6, the focus is on the evaluation task *i.e.* argument validation, submission and preference assignment. §7 presents the wastewater management scenario whereas §8 depicts an example of this scenario. Finally, in §9 the main conclusions are given.

2. The Wastewater Treatment Plant Scenario

In industrialized areas where industrial discharges are connected to the sewer system and finally treated by the WWTP (together with domestic wastewater and rainfall), industrial discharges represent an important load contribution to the Urban Wastewater System (UWS). Several types of industrial discharges with different characteristics (*e.g.* content of organic matter, nutrients and/or presence of pollutants) can affect the growth of micro-organisms into the WWTP and so the WWTP treatment operation and the final result. For that reason, there exists an important body of research intended to improve and increase the knowledge on WWTP operational problems related to influent discharges (*e.g.* [2], [3], [9], [20], [8]). Typically, the representation of this knowledge, based on on-line and off-line data as well as the experts' heuristics, is organized and formalized by means of decision trees and/or knowledge-based flow diagrams (*e.g.* [14], [12], [4]). That is, the knowledge is organized hierarchically by means of top-down descriptions of the interactions between the different parameters and factors used to solve a problem. This representation allows an easy interpretation of the available knowledge, mostly, in terms of cause-effect relations for a concrete problem.

These approaches typically develop their knowledge exploring common benchmark problems with the result of an increasing understanding of such stereotypical scenarios. However, because of the high diversification of industries (*e.g.* long and short-term variations), it is difficult to define *typical industrial operating conditions* and to abstract from external factors such as weather conditions and/or other urban wastewater discharges. As a result, WWTP managers are left with little decision support when confronting a situation that deviates from these benchmark problems. In particular, it is not easy to alter, on the fly, decision trees in order to adapt them to alternative situations (*e.g.* to express a cause-effect relation among diverse factors commonly treated independently).

Guidelines and regulations do exist to protect WWTP from hazardous industrial discharges that can cause operational problems to the WWTP. Such regulations are currently based on the application of discharge standards to point sources defining the permitted quality of discharged wastewater. The description of these standards is made by means of numerical limits (*i.e.* thresholds) for a set of polluting parameters indicating a concentration and/or load. Such numerical limits are defined independently from the particular situation in which the industrial spill is intended, thus ignoring the WWTP particular state and characteristics as well as any external factor that may affect either the spill or the WWTP. However, the fact is that special circumstances may sometimes motivate either to reject discharges that are under legal limits (*e.g.* the WWTP is overloaded) in order to prevent potential complications or to accept discharges that are above legal thresholds since, for example, weather condition (*e.g.* rain may dilute the concentration of a toxic) permits the WWTP to safely deal with the industrial spill and, in doing so, the use of the infrastructure is optimized.

This situation suggests the need for a more flexible decision support mechanism in order to successfully adapt WWTP operation to influent variability and avoid and/or mitigate operational problems into a WWTP. A decision making process that accounts for the above mentioned available knowledge is needed to adapt each decision to the particular situation.

3. Argumentation

Argumentation theory is a rich interdisciplinary area of research that has recently emerged as one of the most promising paradigms for defeasible reasoning and conflict resolution. Its fundamental appeal resides in the provision of intuitive modular models, whereby:

1. Arguments are defined in some underlying logic (an argument and its claim being the formulae from which the claim is derived in the logic).
2. A binary *attack* relation on arguments is defined, that accounts for domain and logic specific notions of conflict or disagreement.
3. *Preferences* among arguments (based, for example, on their relative strength) are used to obtain a binary defeat relation from the attack relation. For example, if arguments *A1* and *A2* symmetrically attack, but *A2* is preferred to *A1* according to some given preference ordering, then *A2* asymmetrically successfully attacks or *defeats A1*.
4. Given the resulting directed graph of arguments, Dung's seminal 'calculus of opposition' [5] then defines the 'winning' arguments under different semantics.

Argumentation, the process in which arguments are constructed and evaluated in light of their conflict-based interactions with other arguments, is inherently dialectical, where the dialogue is driven by the participants' exchange of arguments. This dialectical nature of argumentation is particularly exploited by the use of Argument Schemes (ASs) and Critical Questions (CQs). As described in the informal logic literature (*e.g.* [18]), ASs are used to classify different types of arguments that embody stereotypical patterns of reasoning. Instantiations of ASs can be seen as providing a justification in favor of the conclusion of the argument. The instantiated scheme (what we term an 'argument') can be questioned (attacked) through posing critical questions associated with the scheme. Each CQ can itself be posed as an attacking argument instantiating a particular AS. This AS is then itself subject to critical questioning. The AS & CQ effectively map out the *relevant* space of argumentation, in the sense that for any argument they identify the valid attacking arguments from those that are logically possible. In that sense they provide a natural basis for structuring argumentation based dialogue protocols (*e.g.* [19, 17, 1]).

In §4, we introduce the argument-based model *ProCLAIM* that provides a principled way for deciding collaboratively whether a safety critical action can be performed without causing any undesirable side effect. The model uses the scenario specific patterns of reasoning *i.e.* AS as a means to elicit all the relevant factors for deciding whether a proposed action can safely be performed. This results in a network of interacting arguments. The given arguments are then evaluated on the basis of their content, on who endorses them, and on the basis of their associated evidential support. The result of this argumentative process is a justification to why the proposed action can or cannot be safely performed.

4. Introducing the ProCLAIM Model

The *ProCLAIM* model is intended to assist developers in extending multi-agent systems so that these extended systems support deliberation dialogues among agents for deciding whether a proposed action is safe [16]. *ProCLAIM* can be regarded as defining a centralized medium through which heterogeneous agents can effectively and efficiently deliberate. This centralized medium is embodied by a Mediator Agent (MA) whose role is to warrant the success of the deliberation process. In particular the MA is assigned four main tasks:

- Guide the participants as to what their legal dialectical moves are at each stage of the deliberation. In particular, what schemes they can insatiate. In this way, the deliberation can be regarded as an argumentative process for eliciting the relevant knowledge from the participants (domain experts) as opposed to defining a strategic dialogue in which a better choice of arguments may better serve the agents' individual goals.
- Decide whether or not the participants' submitted arguments are relevant for the discussion and thus added to the graph of interacting arguments. Arguments, although they may be well formed with respect to the underlying model of argumentation, may be nonsensical or too weak when contextualized in the problem at hand. The MA has to prevent these spurious arguments from disrupting the course of the deliberation.
- Submit additional arguments deemed relevant by guidelines and/or previous similar deliberation, that were not taken into account by the participants of the current deliberation. This way it is ensured that all available knowledge is being accounted for when deciding whether or not to perform a safety-critical action.
- Evaluate the submitted arguments that were accepted in order to propose a solution. This involves resolving the symmetrical attacks between arguments into asymmetrical attacks. Once this is done, Dung's calculus of opposition is applied to identify the winning arguments. Thus, in particular, it is identified whether the proposed action can safely be performed or not.

In order to perform these tasks, the MA references four knowledge resources defined by ProCLAIM, depicted in Figure 1, that are briefly described below:

Argument Scheme Repository (ASR): Encodes the scenario specific argument schemes and their associated critical questions. Referenced by the MA in order to direct the participant agents in the submission and exchange of arguments.

Domain Consented Knowledge (DCK): Encodes   the   scenario's   domain consented knowledge. Referenced by the MA in order to account for the domain's guidelines, regulations or any knowledge that has been commonly agreed upon.

Case-Based Reasoning Engine (CBRE): Stores past cases and the arguments given to justify the final decision. Referenced by the MA in order to evaluate the arguments on an evidential basis.

Argument Source Management (ASM): This   component   manages   the confidence in the participants' knowledge on the domain. It is referenced by the MA in order to bias the strength of the arguments on the basis of the agents that endorse them.

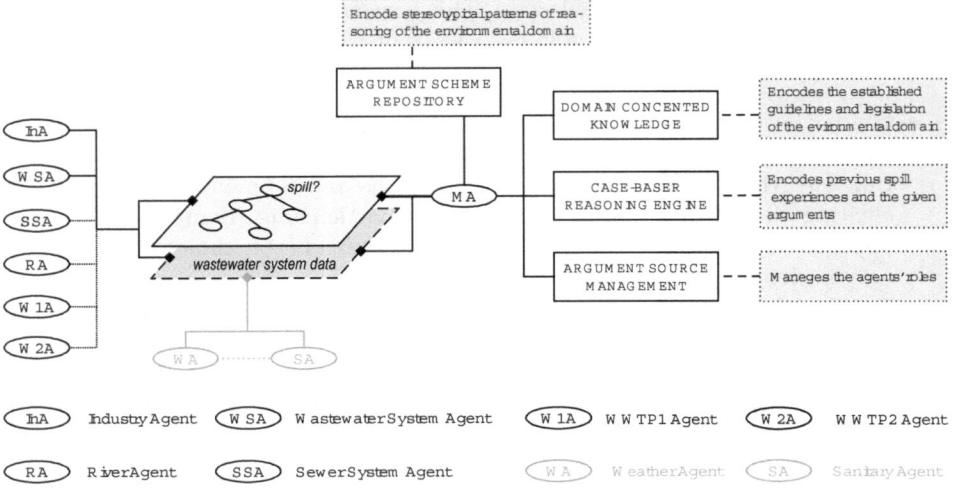

FIGURE 1. *ProCLAIM*'s Architecture. Shaded boxes identify the model's constituent parts specialized for the wastewater management scenario introduced in §7.

ProCLAIM defines two layers of interaction. One in which agents exchange arguments (instantiated schemes) and another one in which they exchange information that is potentially relevant for the deliberation. Thus, for example, in the wastewater management scenario, agents will update each other via the *context layer* on facts such as the industrial spill's content, WWTP's characteristics and the climatological condition. Whereas the argumentation will occur at the *deliberation layer*, see Figure 1. Of course, there may be other required interaction layers for each particular scenario of application, where, for example, agents have to negotiate to decide who does what or to persuade one another on certain issues.

A deliberation in *ProCLAIM* starts with the submission of the argument proposing the initial actions (*e.g.* spill an industrial wastewater discharge). Further submitted arguments attack or defend the justification given for the action proposal. Each submitted argument must instantiate one of the argument schemes in the ASR. Thus, at each stage of the deliberation the *MA* references the ASR in order to indicate the participants those schemes they can instantiate in reply to the already submitted arguments. To prevent spurious arguments, the *MA* validates each of the participants' submitted arguments against the DCK, CBRE and ASM. The *MA* checks that the schemes' instantiations are accepted by the consented knowledge. *MA* also checks whether there is evidence that the submitted argument is a relevant argument and/or whether the agent who submitted the argument is sufficiently trustworthy to exceptionably accept an argument deemed weak by the DCK[1]. In this way, the deliberation is highly focused. Only the reasoning lines defined by the ASR are accounted for, and no spurious argument that may disrupt the course of the deliberation is taken into account.

In parallel, the participant agents update each other the circumstances they are aware of via the *context layer*. Once the graph of interacting arguments is constructed, the *MA* checks whether there are any facts stated to be the case in the *context layer* that was not accounted for by the participant agents. For which the *MA* references the DCK and the CBRE. The *MA* also references these knowledge resources and the ASM in order to assign a preference relation between the mutually attacking arguments to resolve symmetrical attacks into asymmetrical ones. Once this is done, the *MA* identifies the winning arguments by applying Dung's calculus of opposition.

In the following section we describe in more detail *MA*'s task for guiding the agents' argument submission. In §6 we discuss *MA*'s evaluations tasks: validation (§6.1), argument submission (§6.2) and preference assignment (§6.3) to conclude with the proposal of a solution (§6.4)

5. Deliberation as an Argumentative Process for Knowledge Elicitation

The question that *ProCLAIM* aims to resolve is whether or not the proposed action will cause any undesirable side effect that justifies not to perform the action. Thus, the issue is not to motivate the added value of performing the action but rather to identify whether or not the circumstances in which the action is intended are appropriate, *i.e.* that there are no contraindications. To formalize this, let us define the following domains:

R: Domain of facts in circumstances where the action is proposed.

S: Domain of side effects (*i.e.* facts) caused by the action.

A: Domain of actions.

G: Domain of undesirable goals. **G** contains an element nil denoting the absence of undesirable goals.

Let **R**, **A** and **S** be sets of grounded first order predicates and **G** be a set of propositions, so that in the environmental scenario the fact that the industry ind has wastewater ww and it is connected to the WWTP wwtp can be denoted as has_ww(ind,ww), connected(ind,wwtp) $\in$ **R**.

Let us also introduce $\mathfrak{C}_F$ as the set of facts submitted by the participant agents' at the *context layer*. Then, the circumstances in which the deliberation takes place are shaped by the elements, facts, in $\mathfrak{C}_F$, since each and every fact

---

[1]Suppose that a trustworthy agent submits an argument A proposing an alternative action to warrant the safety of the initially proposed action. Both, the DCK and CBRE may deem A too weak to be accepted. However, because the agent is trustworthy, argument A may exceptionally be accepted. That is, added to the graph of interacting arguments.

the participant agents deem potentially relevant for the decision making should be added to $\mathfrak{C}_F$. Note that, in particular, $\mathfrak{C}_F \subseteq \mathbf{R}$. Hence, under this formalization, the main question *ProCLAIM* addresses can be rephrased as: Accounting for the possible complementary courses of action in $\mathbf{A}$, are there $r_1, ..., r_n \in \mathfrak{C}_F$ because of which the proposed action will bring into being a side effect $s \in \mathbf{S}$ that realizes an undesirable goal $g^- \in \mathbf{G}^- - \{\texttt{nil}\}$ that justifies not to perform the proposed action?

To address the above question, *ProCLAIM* defines an argumentative process formalized in terms of a structured set of AS & CQ. These schemes and critical questions conform a protocol-based exchange of arguments that allows the identification of arguments that can be submitted at each stage of the deliberation. This protocol is used by the *MA* in order to guide the participant agents in their argument submission. Through this guidance, participant agents (experts) are led to unfold the relevant facts in $\mathfrak{C}_F$ and complementary actions in $\mathbf{A}$ and indicate why they are relevant for the decision making.

ProCLAIM defines ASs and their associated CQs on two levels of abstraction. At the first more abstract level, the schemes – although tailored for deliberating over the safety of an action – abstract from any particular scenario. These ASs & CQs conform *ProCLAIM*'s basic protocol-based exchange of arguments. For each particular application these more abstract schemes are then specialized for the intended application. These scenario-specific schemes conform the Argument Scheme Repository.

In the following, in §5.1 *ProCLAIM*'s basic protocol-based exchange of arguments is introduced and in §5.2 the illocutions defined by *ProCLAIM* for the participants interaction are described.

5.1. ProCLAIM's basic protocol-based exchange of arguments

An argument in *ProCLAIM* expresses a causal relation among elements of the sets $\mathbf{R}$, $\mathbf{A}$, $\mathbf{S}$ and $\mathbf{G}$. So that if $R \subseteq \mathfrak{C}_F$, $A \subseteq \mathbf{A}$, $S \subseteq \mathbf{S}$ and $g \in \mathbf{G}^-$ then, if $g \neq \texttt{nil}$, the tuple $< R, A, S, g >$ indicates that in the selected context of facts R the actions A will bring into being the side effects S that realize the undesirable goal g. While, if $g = \texttt{nil}$, that is, $< R, A, S, \texttt{nil} >$ the relation indicates that in the selected context of facts, the actions are not expected to cause any severe undesirable side effect.

A deliberation starts with the submission of an argument proposing the initial action (*e.g.* $\texttt{spill(ind,ww,wwtp)}$) with the assumption that there are no contraindications for its performance. However, in order to propose the initial action, a minimum context of facts must hold. In the environmental scenario, these preconditions would be that an industry $\texttt{ind}$ has wastewater ($\texttt{has\_ww(ind,ww)}$) and is connected to the WWTP $\texttt{wwtp}$ ($\texttt{connected(ind,wwtp)}$). So, in general, if the action proposal is $a_0 \in \mathbf{A}$ and the minimum context of facts is $Cmin \subseteq \mathfrak{C}_F$, then the argument expresses the relation $< Cmin, \{a_0\}, \{\}, nil >$. In the environmental scenario it would be:

$< \{\texttt{has\_ww(ind,ww)}, \texttt{connected(ind,wwtp)}\}, \{\texttt{spill(ind,ww,wwtp)}\}, \{\}, nil >$.

In the course of the deliberation agents will submit further arguments that attack or defend the assumption that no contraindications exists. These arguments will introduce new facts, from $\mathfrak{C}_F$, or new complementary actions, from $\mathbf{A}$, and thus, they will extend the contexts of facts and actions. To clearly distinguish between the new introduced facts or actions in an argument, let us extend the 4-tuple in to a 6-tuple:

$$< \mathcal{C}, R, \mathcal{A}, A_{comp}, S, g >$$

where:

$\mathcal{C}$: is the context of facts already introduced. $\mathcal{C}$ contains the minimum set of facts required for proposing the initial action and $\mathcal{C} \subseteq \mathfrak{C}_F$.

R: R is the newly introduced set of facts, $R \subseteq \mathfrak{C}_F$ and $R \cap \mathcal{C} = \emptyset$.

$\mathcal{A}$: is the set of actions intended to perform, $\mathcal{A} \subseteq \mathbf{A}$ and $\mathcal{A}$ contains the proposed action under debate.

A_{comp}: is a set of complementary actions proposed to prevent or mitigate an undesirable side effect. $A_{comp} \subseteq \mathbf{A}$ and $A_{comp} \cap \mathcal{A} = \emptyset$.

S: is a set of side effects, $S \subseteq \mathbf{S}$.

g: is an undesirable goal $g \in \mathbf{G}^-$, possibly $\texttt{nil}$.

For different arrangements of values of the above elements we obtain different ASs. As discussed above, the deliberation must start with $\mathcal{A}$ being the proposed action and $\mathcal{C}$ being the minimum required context for proposing the action. The first argument scheme is $AS1$ that an agent instantiates to start the deliberation (see Figure 2). So if:

$\mathcal{A} :=$ proposed action, and
$\mathcal{C} :=$ minimum context

AS1: $< \mathcal{C}, \{\}, \mathcal{A}, \{\}, \{\}, \texttt{nil} >$; with the narrative version being:

AS1:
In circumstances $\mathcal{C}$,
the proposed course of action $\mathcal{A}$ can safely be performed.

Associated to $AS1$ is the critical question:

AS1_CQ1: Is there a contraindication for performing the proposed action?

Linked to this CQ is the scheme $AS2$ that defines arguments that introduce a contraindication. Let $R \subseteq \mathfrak{C}_F$, $S \subseteq \mathbf{S}$ and $g \in \mathbf{G}^-\{\texttt{nil}\}$, then:

AS2: $< \mathcal{C}, R, \mathcal{A}, \{\}, S, g >$.
In circumstances $\mathcal{C}$,
because R holds, actions $\mathcal{A}$ will cause a side effect S
that will realize some undesirable goal g.

There are three CQs associated to *AS2*. Each of these CQs has, in turn, an associated AS that embodies the question as an attacking argument. Note than in reply to an argument instantiating *AS2*, the context of facts is extended with the set of facts R. Namely, $\mathcal{C} := \mathcal{C} \cup R$.

AS2_CQ1: : Are the current circumstances such that the stated effect will not be achieved?

> **AS3:** $< \mathcal{C}, R, \mathcal{A}, \{\}, \{\}, \texttt{nil} >$.
> In circumstances $\mathcal{C}$,
> because R holds, the side effect S is not expected as caused by $\mathcal{A}$.

AS2_CQ2: : Are the current circumstances such that the achieved effect S will not realize the stated goal g?

> **AS4:** $< \mathcal{C}, R, \mathcal{A}, \{\}, S, \texttt{nil} >$.
> In circumstances $\mathcal{C}$,
> and assuming $\mathcal{A}$ will be performed,
> it is because of the fact that R holds that S does not realizes **g**

AS2_CQ3: : Is there a complementary course of action that prevents the achievement of the stated effect S?

> **AS5:** $< \mathcal{C}, \{\}, \mathcal{A}, A_{comp}, \{\}, \texttt{nil} >$.
> In circumstances $\mathcal{C}$,
> the complementary course of action A_{comp}
> prevents actions $\mathcal{A}$ from causing the side effect S.

As depicted in Figure 2, arguments that instantiate schemes *AS3*, *AS4* or *AS5* can be attacked by arguments that instantiate either scheme *AS2* again, or, by arguments that instantiate scheme *AS6*:

AS3_CQ1, AS4_CQ1, AS5_CQ1:: Are the introduced factors (facts or actions) themselves a contraindication for performing $\mathcal{A}$ in circumstances $\mathcal{C}$?

> $\mathcal{C} := \mathcal{C} \cup R$;
> $\mathcal{A} := \mathcal{A} \cup A_{comp}$;
> **AS6:** $< \mathcal{C}, \{\}, \mathcal{A}, \{\}, S2, g >$.
> In circumstances $\mathcal{C}$,
> $\mathcal{A}$ will cause the side effect $S2$
> that realizes the undesirable goal g.

AS3_CQ2, AS4_CQ2, AS5_CQ2,: Is there a contraindication for performing the proposed action?

This CQ links again the argument scheme *AS2*. Further, as depicted in Figure 2, to *AS6* are associated the same CQs as to *AS2*.

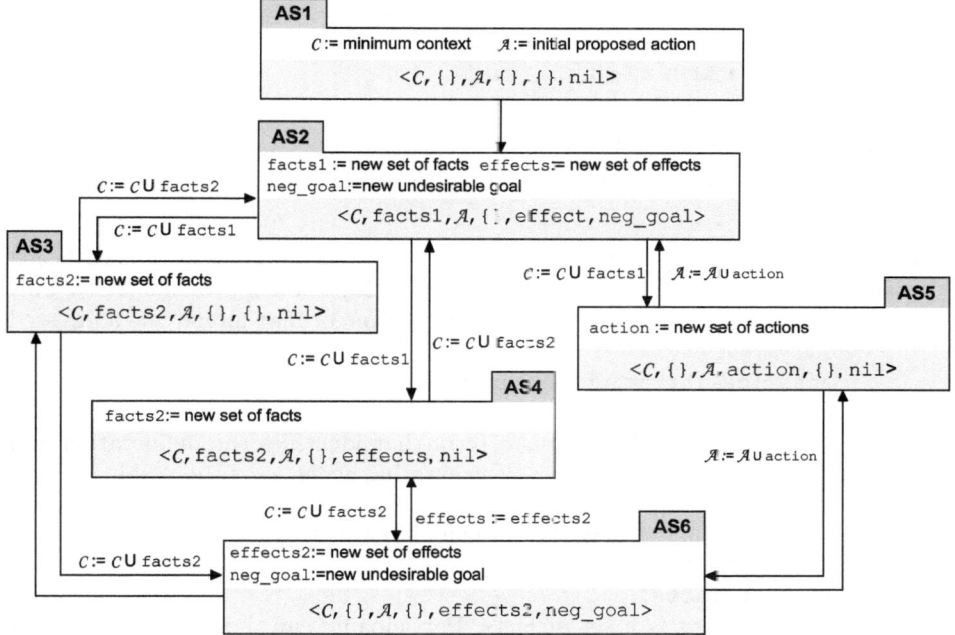

FIGURE 2. Argument schemes' interaction via their associated critical questions.

The above introduced circuit of schemes linked via their associated CQs define a protocol-based exchange of arguments specialized for deliberating whether a proposed action will or will not cause an undesirable side effect. Despite the schemes' specialization, it requires some skill in argumentation to effectively and efficiently instantiate these schemes on the fly. As a result, the success of a deliberation is, to some extent, limited by the players argumentation ability rather than by their knowledge of the problem at hand. To overcome this problem, *ProCLAIM* defines a lower layer of abstraction of the above six ASs. These more specific ASs are not only tailored for arguing over an action's safety, but they are specialized for the deliberation in a particular scenario. That is, they aim to capture the scenario's stereotypical reasoning patterns. Example of these scenario specific schemes are given in §8.

These scenario specific schemes are transparent for experts in the domain and thus can be instantiated with no overhead. These specific ASs and their associated CQs conform the ASR. That is, these are the schemes which the *MA* uses to guide the participants in their argument submission.

The construction of the ASR thus consists in semi-instantiating the above six abstract ASs. We are currently developing a methodology to facilitate this task which otherwise is rather ad-hoc.

5.2. Agent's Interaction Protocol

In what follows we overview the illocutions defined by *ProCLAIM* for the agents interaction:

- Enter and leave the deliberation. In entering, the *MA* has to identify each agent *w.r.t* the ASM (the agent role, the institution it represents, *etc.*). This of course may be included in the `enter` locution, or may already be addressed previously.
  - `enter(deliberationID, agentID)`
  - `leave(deliberationID)`

  Where `deliberationID` is a term that identifies the deliberation and `agentID` is a term that identifies the entering agent *w.r.t* the ASM.
- Participants continuously update the set of facts $\mathfrak{C}_F$. This involves introducing facts as well as retracting from facts later shown to be false.
  - `assert(facts)`
  - `retract(facts)`

  Where `facts` is a set of facts. Participants can only assert facts that are consistent with those in $\mathfrak{C}_F$. And obviously they can only retract those facts that are in $\mathfrak{C}_F$. Again, particular applications may define preconditions for asserting or retracting facts.
- As the set of facts $\mathfrak{C}_F$ is updated, the *MA* updates the participants with the new set of facts by broadcasting a message:
  - `circumstances(facts)`

  With `facts` is the updated $\mathfrak{C}_F$. In some contexts not all participants should have complete access to all the available facts. In each particular case, the views of each participant on $\mathfrak{C}_F$ must be defined.
- Participants submit arguments legal *w.r.t* the ASR.
  - `move(ID,argue(argument),target)`

  Where `ID` is the identifier of the submitted argument, grounded to the actual value if the move is accepted; and `target` is the identifier of the move it replies to.
- The *MA* broadcasts each change in the tree of argument to all participants. One way to do this is to send the updated tree of arguments `tree` to the participants:
  - `madeMoves(tree)`
- During the deliberation participants can express which arguments they endorse, which they do not and which they disagree with:
  - `agree(id)`
  - `disagree(id)`
  - `abstain(id)`

Where `id` is the identifier of an argument in the tree of arguments. By agreeing or disagreeing participants may bias the strength of the submitted arguments. A participant may disagree with an argument herself submitted and agree with an argument that attacks an argument she submitted. If she agrees with two conflicting arguments, she should *abstain*.

- Although participants may have a copy of the ASR (and any other resource specific for the application) and so may be able to reason what the legal deliberation moves (arguments and challenges) are at each stage of the deliberation, with any change in the tree of arguments the *MA* broadcasts to all participants their updated legal moves:
  - `legalMoves(moves)`

  Where `moves` is a list of legal deliberation moves of the form `move(ID, move, target)`. The `target` indicates the element of the tree of argument this move, if submitted, would reply to. `move` is an argue move `argue(argument)` with `argument` is a semi-instantiated scheme.

- When participants have no more deliberation moves to make, then they send the message:
  - `noMoreArguments()`

- The *MA* may conclude the deliberation at any time:
  - `end(solution,reason))`

  Proposing a `solution` to the deliberation and providing the `reason` for terminating the deliberation (*e.g.* timeout or all participants expressed that they have no more arguments).

These locutions compose a basic protocol that allows an agent to undertake a *ProCLAIM* deliberation. Participant interaction is via the *MA* who works also as a gatekeeper[2]. Participants request to enter, leave, submit an argument, *etc.*, so this dialogue locutions can be wrapped into a *request-inform* loop, where participants request to submit an argument (`request(move(ID, argument,target))`), and if the requested locution is accepted, the *MA* informs the sender that the locution was accepted (and in the argue locution case, the *MA* will broadcast the new state of the tree of arguments). If the locution is not accepted, the *MA* will provide the sender with the reason for its rejection.

Natural extensions to the protocol are requests from the participants for the legal moves, tree of arguments and current circumstances. So if, for any reason, a participant looses communication he can explicitly request for an update of the deliberation stage and so not to wait for a broadcast.

6. *MA*'s Evaluative Tasks

In this section we describe *MA*'s evaluative tasks which encompasses (1) validating that the submitted arguments are relevant for the deliberation, (2) checking that

---

[2]Of course, one may decide to split the *MA* into a number of agents all doing different tasks: gatekeeper, legal moves provider, argument validation, evaluation *w.r.t* DCK, CBRE and ASM.

no relevant factor was left unaddressed by the participants, and (3) assign a preference relation among the mutually attacking arguments to identify the winning arguments.

6.1. Argument Validation

For a deliberation move to be accepted in *ProCLAIM* , it has to pass a validation process that involves:

- Unless it is the first move, in which case it has to instantiate an *AS1* scheme specific for the domain, the move has to be a legal reply, *w.r.t* the ASR, to another deliberation move in the tree of arguments.
- The submitted argument has to be validated by *ProCLAIM*'s knowledge resources (*i.e.* DCK, CBRE and ASM) to prevent spurious or too weak arguments in the tree of arguments that may unnecessarily populate the tree of arguments. If the argument is validated, then it is added. However an accepted argument, although weak, may be dismissed during the deliberation if other participants reject it.

To accept an argument as valid is to consider it as sensible, not necessarily stronger than the argument it replies to. Thus, it is a measure to prevent inclusion of non-interesting arguments (spurious or too weak) in the deliberation, and hence helps focusing the deliberation on the relevant matters to be discussed. The idea is that if the domain knowledge validates an argument, the argument is accepted. If not, the argument may still be accepted because of the trust in the submitter expertise on the domain area the argument relates to; and/or because there are records of successful uses of the submitted argument.

6.2. Argument Submission

After the participants have submitted their arguments, the *MA* checks whether according to the DCK there are facts in the $\mathfrak{C}_F$ or possible intervention plans (complementary courses of actions) known to prevent some undesirable state that are not mentioned by the participants. If this is the case, these factors are introduced in the deliberation by the *MA* as an additional argument or arguments.

A similar idea is taken with the CBRE: Arguments that were not submitted by the participants nor by the DCK, but were submitted in previous similar deliberation and are relevant for the current deliberation. For example, if in previous deliberations a novel procedure for preventing an undesirable state was successfully proposed and later successfully performed, this procedure may be unknown to the participants and has not yet been integrated into the DCK. So this novel procedure may be proposed as an argument by the CBRE.

Thus, a decision is taken that accounts for the participant agents (experts), the domain consented knowledge and the empirical evidence.

6.3. Arguments' Preference Assignment

What remains to be done is to resolve the mutual attacks between arguments with Dung's calculus of opposition to determine whether the proposed action is safe or not. While the tree of arguments maps out all the relevant factors for the decision, arranging them in a convenient way, *ProCLAIM*'s knowledge resources propose a solution on the basis of these highlighted factors. Where a solution is proposed by resolving symmetrical attacks between arguments into asymmetrical attacks, that is, assigning a preference between mutually attacking arguments. *ProCLAIM* defines three knowledge resources from which to derive the preference assignment between arguments:

1. The **DCK** provides a preference assignment on the basis of the domain consented knowledge, that is guidelines, standard criteria, standard intervention plans, *etc.*.
2. The **CBRE** provides a preference assignment based on evidence gathered from previous similar cases. If in previous similar cases an argument, say, *A1* usually defeats argument *A2*, the CBRE will suggest *A1* to be preferred to *A2*. The CBRE is described in detail in [15].
3. The **ASM** provides a preference assignment based on the trust in the experts endorsing one or another argument. So that if *A1* and *A2* are two mutually attacking arguments and the agents endorsing *A2* are more trustworthy than those endorsing *A1*, then the ASM will suggest that *A2* should be preferred to *A1*.

6.4. Proposing a Solution

Each of these three preference assignments embodies a different and independent perspective from which a proposal can be evaluated. Thus, rather than proposing a formula that flattens the three preference assignments into one, the idea is to provide a qualitative solution. In this solution, the DCK assignment is taken as the central assignment and the other two are in its support or against it. Thus, if the arguments *A1* and *A2* mutually attack each other, a qualitative solution is of the kind:

DCK *weakly* suggests preferring *A1* to *A2*.
But ASM *strongly* suggests preferring *A2* to *A1*.
And CBRE *weakly* supports ASM's assignment.

The decision whether *A1* should defeat *A2* or the other way around is scenario-dependent. Note that if all three assignments are in agreement, or at least, they are not in disagreement; a decision can be taken automatically. However, in case of disagreement it would in general require the intervention of a human expert to take the final decision. Thus, whenever all the assignments in a tree are in agreement, the final decision, whether to accept or reject a proposed action, can be automatized. And whenever a disagreement occurs, a warning is triggered that identifies the source of the conflict and the assessments (preference assignments) of each of the independent knowledge resources.

7. The Industrial Wastewater Scenario

This section presents the wastewater management context in which industrial discharges are released. The wastewater scenario is reduced on purpose to ease the understanding of the argument-based methodology when dealing with critical decisions into the wastewater system (*e.g.* the safety of an industrial discharge containing a polluting substance).

Often industries deal with the produced wastewater by connecting to the sewer system. Therefore, industries can be considered part of the Urban Wastewater System (UWS) whose main components are shown in Figure 3. We consider every relevant element as a software agent, *i.e.* an autonomous entity that can interact with other ones in order to achieve an individual or common goal [13], [21]. In this specific example, we consider the following three proponent agents that can significantly participate in the argument-based deliberation when dealing with an industrial wastewater discharge into the UWS:

- **Industry Agent (IA)**: represents individual industries and/or groups of industries that need to manage their produced wastewater as a result of their production process. *IA* discharge their produced wastewater into the sewer system, where it is collected together with other inflows and transported to the WWTP (from now this course of action is called a_0).
- **Wastewater Treatment Agent Manager (WTAM)**: represents the manager of WWTP. Its main function is to keep track of wastewater flow arriving at WWTP as well as to supervise and control the treatment process. It gives convenient alarms when necessary and the orders to change the operational set points.
- **River Consortium Agent (RCA)**: represents the maximum authority in the catchment, whose main objective is to preserve the river quality. Its main functions are to manage and coordinate a group of WWTPs in the river catchment as well as to monitor river quality and to prevent possible hazardous contamination by supervising *IA* and *WTAM*.

All of these agents, representing experts in the wastewater treatment domain, can take part in the deliberation process having different degrees of responsibility and making use of their expertise in front of safety critical decisions. In Figure 1, we show all these agents in the context of *ProCLAIM* in this scenario. In order to illustrate the evaluation of the arguments posed by these experts, it is important to know the acquaintances among them and the degree of confidence in their arguments. Figure 4 shows the main relations among the three agents considered in this specific example. The number in the box indicates the order of the agents' reputation specific for each type of discharge. *WTAM* knows the operation of its WWTP and thus the degree of confidence of his/her arguments is high. However, the figure shows that when dealing with discharges containing priority polluting substances, the *CRA* is the first on the reputation scale so that their argument evaluation will be readjusted to higher strength.

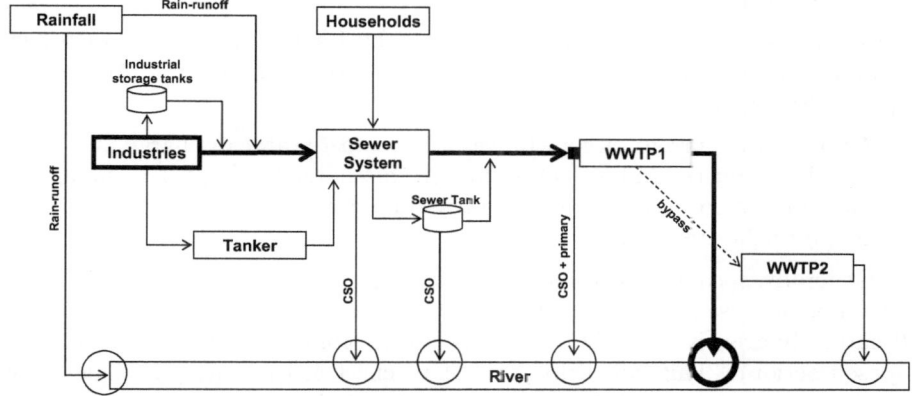

FIGURE 3. Urban Wastewater System (UWS). In **bold** the normal path followed by an industrial discharge. CSO: Combined Sewer Overflow.

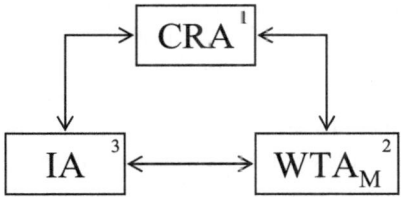

FIGURE 4. Acquaintances of agents and reputation (indicated by the numbers inside the boxes) in the proposed wastewater scenario.

All of these above mentioned considerations will be taken into account by the *MA* to evaluate the arguments. In this process, we want to ensure that all the legal moves in the deliberation process are posed and check whether the participants submitted arguments should be accepted. If accepted they will be part of the argument graph for a specific deliberation process, such as the one depicted in Figure 5.

8. A Running Example

Let us suppose that an industry, represented by its IA, proposes a wastewater discharge claiming that no undesirable effects will occur afterwards. Accordingly, IA poses the argument $Arg1$:

> **Arg1**: In the current circumstances (*i.e.* a wastewater discharge and a WWTP) industry Ind_i will effectuate the discharge (action $\mathbf{a_0}$), claiming that this action ($\mathbf{a_0}$) will not cause any side effect S, *i.e.* any undesirable goal g to the treatment system.

Generally speaking, when an industry agent (IA) claims to discharge its wastewater because no negative effects occur (*e.g.* $Arg1$), a critical question that will naturally arise is "AS1_CQ1: Is there a contraindication for undertaking the proposed action?" This will help the MA to check if the following dialog move is legal. Assuming that $WTAM$ knows that the discharge contains cadmium and believes cadmium is a contraindication for the treatment process, because it can provoke dispersed growth due to the inhibition of Extracellular Polymeric Substances (EPS) ([6], [7]), $WTAM$ submits $Arg2$:

> **Arg2**: If in current circumstances industry Ind_i effectuate the discharge ($\mathbf{a_0}$) containing cadmium ($\mathbf{r_4}$), this will cause EPS inhibition ($\mathbf{s_3}$) and hence provoke dispersed growth ($\mathbf{g_4}$).

Arg2 introduces new important information about the discharge (*i.e.* the discharge contains cadmium that can cause EPS inhibition). Different experts on the domain can naturally start a dialogue of attacking and supporting arguments, seeking for more information, for alternative actions, *etc.* to finally decide on the possible actions to be taken to prevent WWTP problems. According to this, other possible counterarguments exist that are risen by three new critical questions (they are meant to limit the possible counterarguments, discarding the ones not relevant for the discussion, and hence look for the key information):

> **AS2_CQ1**: Are the current circumstances such that the stated effect will be achieved? That is equivalent, in the presented example, to question whether the concentration of the cadmium, given the current circumstances, is enough to produce the undesirable effect (*i.e.* EPS inhibition) even if it is under legal thresholds (*e.g.* rain can dilute cadmium concentration).

> **AS2_CQ2**: Are the current circumstances such that the achieved effect will realize the stated negative goal? That is, to explore other relevant circumstances in the context that makes the negative goal nil (*e.g.* synergetic effects with other pollutants, *etc.*).

> **AS2_CQ3**: Is there a course of action that prevents the achievement of the stated effect, that is, to explore the possible actions that can prevent or mitigate the negative effect (*e.g.* increase dissolved oxygen, decrease wasted activated sludge, *etc.*).

Figure 5 shows some of the possible lines of reasoning when dealing with the industrial discharge containing cadmium. Following the example, **AS2_CQ1** and **AS_CQ3** pose an attack to **Arg2**. Consequently, **Arg3** and **Arg4** (see the table in Figure 5) attack **Arg2** (*e.g.* they are instances with new information about the discharge or an alternative action, respectively).

Arg3: If in current circumstances industry Ind_i effectuates the discharge (a_0) containing Cadmium (r_4) it will not cause EPS inhibition (s_3), hence it does not provoke dispersed growth (g_4) since it is raining and the rain (r_5) will dilute the discharge.

Arg4: If in current circumstances industry Ind_i effectuates the discharge (a_0) containing cadmium (r_4), it will not cause EPS inhibition (s_3), hence it does not provoke dispersed growth (g_4) since it can be added coagulants/flocculants (a_3) to help the settling process.

Notice that no agent attacks **Arg2** in response to **AS2_CQ2**. This means that, as far as it is known, no argument exists that claims that – given the negative effect of EPS inhibition – there is no negative goal associated with it. However, these aforementioned factors (*i.e.* rain and addition of coagulants/flocculants) can provoke some secondary effects which are counterarguments for **Arg3** and **Arg4**:

Arg5: If in current circumstances, industry Ind_i effectuates the discharge (a_0) containing cadmium (r_4), the rain (r_5) can dilute the discharge, thus preventing EPS inhibition (s_3) and dispersed growth (g_4). However, considering its intensity, it can cause a hydraulic shock at the WWTP and provoke a washout of activated biomass.

Arg6: If in current circumstances, industry Ind_i effectuates the discharge (a_0) containing cadmium (r_4), the addition of coagulants/flocculants can mitigate EPS inhibition (s_3), and hence prevent dispersed growth (g_4). However, if an overdose is applied, there will be an episode of aquatic life toxicity in rivers where the WWTP effluent is discharged.

Arg7: If in current circumstances, industry Ind_i effectuates the discharge (a_0) containing cadmium (r_4), the addition of coagulants/flocculants can mitigate EPS inhibition (s_3), and hence preventing dispersed growth (g_4). However, if an overdose is applied, there will be a complete charge reversal and this will re-stabilize the colloid complex and cause settler problems.

Once the argument graph is constructed, the *MA* has to determine the winning arguments. In this example (see Figure 5) we are going to consider the following: It is not raining, thus the line of reasoning on the left of the argument graph is automatically discarded; and there are no past experiences registered. Therefore, the conflict between **Arg4** and **Arg6** needs to be solved, that is, whether the action proposed to mitigate the effect of the toxic is acceptable or not. A similar procedure should be started to resolve the conflict between **Arg4** and **Arg7**. On the basis of the DCK (articulate in terms of *R*, *A*, *S* and *G*) and the reputation

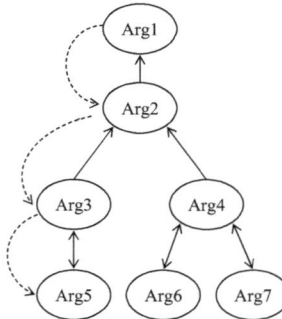

ID	Argument	Proponent Agent
Arg1	<R(ww, wwtp), A(a$_0$), S( ), G'(nil)>	IA
Arg2	<R(ww(r$_4$), wwtp), A(a$_0$), S(s$_3$), G'(g$_4$)>	WTA$_M$
Arg3	<R(ww(r$_4$), wwtp, **rain**), A(a$_0$), S( ), G'(nil)>	IA
Arg4	<R(ww(r$_4$), wwtp), A(a$_0$, a$_3$), S( ), G'(nil)>	IA
Arg5	<R(ww(r$_4$), wwtp, **rain**), A(a$_0$), S(s$_5$), G'(g$_7$)>	WTA$_M$
Arg6	<R(ww(r$_4$), wwtp), A(a$_0$,a$_3$), S(s$_6$), G'(g$_8$)>	CRA
Arg7	<R(ww(r$_4$), wwtp), A(a$_0$,a$_3$), S(s$_6$), G'(g$_9$)>	WTA$_M$

Attack relation	⟶
Dialog move	---⟶

FIGURE 5. Argument graph that captures the moves in a dialog over the acceptability of a toxic industrial discharge into the WWTP. Each node of the tree holds one argument described in the table. Each new introduced factor is highlighted in **bold**.

of the agents involved, different strengths can be given to each of the arguments in order to finally identify the winner and which course of action is the safest for the actual WWTP performance.

Accordingly, *IA* holds with **Arg4** since it believes that it is possible to mitigate the problem by adding coagulants/flocculants. However, as depicted in Figure 4, the idea is that both *WTAM* and *CRA* have a higher reputation and their arguments are ranked higher than *IA*'s ones. However, when dealing with priority substances, *CRA* is the maximum authority which means that **Arg6** has a higher level of confidence than **Arg7**. The aim of the overall system is to preserve the river water quality by allowing the agents to participate into the deliberation and to finally take the safest environmental decision. In a way, this is a Group Decision Support System that is based on an argumentation framework ([10]).

For this specific case, and without considering past experiences, the discharge is considered unsafe. Although a mitigating action can avoid operational problems at the WWTP, there is the Consortium River Agent (CRA) claiming that this action can have a pervasive effect for the aquatic life once the WWTP effluent reached the river (*e.g.* reported by [11], waterborne cadmium can cause severe, acute toxicological and physiological effects to aquatic organisms). Obviously, the toxicological effect will depend on the species, dose, exposure, chemical mixtures, *etc.*, so that from here new relevant reasoning lines could to be studied.

From now, since the discharge proposed by the *IA* should be rejected for the present circumstances, another course of action needs to be considered to manage the discharge (*e.g.* specific pre-treatment at industry, store the discharge – if storage tanks are available – until the system is in proper conditions to hold the discharge, and/or any other possible action that could increase the argument graph for this specific problem). Moreover, since the action proposed by the *IA* is rejected and considering that it claimed safety, its reliability (in terms of the notion of reputation) will diminish.

9. Conclusions

In this article we have presented an argument-based model, *ProCLAIM*, intended to facilitate the implementation of effective and efficient deliberation about safety-critical actions. For the practical realization of such deliberation, the model features a number of resources managed by a Mediator Agent, such as a repository of argument schemes that encodes the scenario's application stereotypical reasoning patterns (the ASR), a knowledge base that encodes the domain consented knowledge (the DCK), a Case-Based Reasoning engine (CBRE) that allows accounting for previous similar deliberation and a knowledge resource that encodes the degree in which participants are to be trusted regarding the matters under discussion (the ASM).

In this article we exemplify the use of *ProCLAIM* in a complex environmental scenario. Industrial wastewater discharges represent a main concern for WWTP managers. The variability of possible industrial discharges, the complex and often uncertain knowledge and information related to the activated sludge based processes to treat wastewater, make the management of industrial discharges both a challenge and a problem. It is of special importance to use timely and precise information to understand and make decisions about the stated problem, as well as, to develop criteria for evaluating the possible solutions for each situation. The understanding of the problem, which focuses on the negative side effects, in order to prevent or mitigate them, makes it less difficult to explore the current context and possible actions, and thus to articulate the problem beyond numerical thresholds. The argumentative approach for decision support systems to solve environmental problems is just spreading out and challenging other more traditional knowledge-based approaches. *ProCLAIM* provides environmental engineers with a tool that provides support to decision makers in non-standard situations. This article proposes a different way to conceptualize the decision making process in order to offer a reliable source of understanding the problems that may arise in the WWTP, jointly with possible solutions (*e.g.* alternative actions). This new conceptualization allows to proceed beside legislation while also taking into account the actual state of the plant and other relevant factors in order to make a fully informed decision.

9.1. Future Work

Currently we are working on a methodology for constructing the *ProCLAIM*'s Argument Scheme Repository. The key idea behind this attempt is to further specialize the introduced argument schemes to construct a repository of domain dependent schemes.

Acknowledgment

The authors would like to acknowledge support from the EU funded project *SHARE-it*: *Supported Human Autonomy for Recovery and Enhancement of Cognitive and Motor Abilities using Information Technologies* (FP6-IST-045088). The views expressed in this article are not necessarily those of the *SHARE-it* consortium. The authors would like to acknowledge the anonymous reviewers for their work in reviewing earlier versions of this paper and their valuable comments.

References

1. K. Atkinson, T. Bench-Capon, and P. McBurney, *A Dialogue Game Protocol for Multi-agent Argument over Proposals for Action*, Argumentation in Multi-Agent Systems: First International Workshop, ArgMAS 2004, New York, NY, USA, July 19, 2004: Revised Selected and Invited Papers (2005).

2. J. Comas, I. Rodríguez-Roda, K.V. Gernaey, C. Rosen, U. Jeppsson, and M. Poch, *Risk assessment modelling of microbiology-related solids separation problems in activated sludge systems*, Environmental Modelling and Software **23** (2008), no. 10–11, 1250–1261.

3. J. Comas, I. Rodríguez-Roda, M. Sànchez-Marrè, U. Cortés, A. Freixó, J. Arráez, and M. Poch, *A knowledge-based approach to the deflocculation problem: integrating on-line, off-line, and heuristic information*, Water Research **37** (2003), no. 10, 2377–2387.

4. U. Cortés, M. Martínez, J. Comas, M. Sànchez-Marrè, M. Poch, and I. Rodríguez-Roda, *A conceptual model to facilitate knowledge sharing for bulking solving in wastewater treatment plants*, AI Communications **16** (2003), no. 4, 279–289.

5. P.M. Dung, *On the acceptability of arguments and its fundamental role in nonmonotonic reasoning, logic programming and n-person games*, Artificial Intelligence **77** (1995), no. 2, 321–357.

6. I.D.S. Henriques, R.D. Holbrook, R.T. Kelly, and N.G. Love, *The impact of floc size on respiration inhibition by soluble toxicantsa comparative investigation*, Water Research **39** (2005), no. 12, 2559–2568.

7. I.D.S. Henriques and N.G. Love, *The role of extracellular polymeric substances in the toxicity response of activated sludge bacteria to chemical toxins*, Water Research **41** (2007), no. 18, 4177–4185.

8. M. Henze, R. Dupont, P. Grau, and A. de la Sota, *Rising sludge in secondary settlers due to denitrification*, Water Research **27** (1993), no. 2, 231–236.

9. D. Jenkins, M.G. Richard, and G.T. Daigger, *Manual on the Causes and Control of Activated Sludge Bulking, Foaming, and Other Solids Separation Problems*, third ed., IWA Publishing, London, UK, 2003.

10. N. Karacapilidis and D. Papadias, *Computer supported argumentation and collaborative decision making: the HERMES system*, Information Systems **26** (2001), no. 4, 259–277.

11. C.A. Mebane, *Cadmium Risks to Freshwater Life: Derivation and Validation of Low-Effect Criteria Values using Laboratory and Field Studies*, Tech. report, U.S. Geological Survey Scientific Investigations Report 2006-5245, 2006.

12. I. Rodríguez-Roda, M. Sànchez-Marrè, J. Comas, J. Baeza, J. Colprim, J. Lafuente, U. Cortés, and M. Poch, *A hybrid supervisory system to support WWTP operation: implementation and validation*, Water Science & Technology **45** (2002), no. 4, 289–297.

13. S.J. Russell and P. Norvig, *Artificial intelligence: A modern approach*, 2nd ed., Pearson Education, Inc., 2003.

14. P. Serra, M. Sànchez-Marreè, J. Lafuente, U. Cortés, and M. Poch, *ISCWAP: A knowledge-based system for supervising activated sludge processes*, Computers and Chemical Engineering **21** (1997), no. 2, 211–221.

15. P. Tolchinksy, S. Modgil, U. Cortés, and M. Sànchez-Marrè, *Cbr and argument schemes for collaborative decision making*, Proc. of Confe. on Computational Models of Argument (COMMA) (P.E. Dunne and T.J.M. Bench-Capon, eds.), Frontiers in Artificial Intelligence and Aplications, vol. 144, IOS Press, 2006, pp. 71–92.

16. P. Tolchinsky, U. Cortés, S. Modgil, F. Caballero, and A. López-Navidad, *Increasing Human-Organ Transplant Availability: Argumentation-Based Agent Deliberation*, Intelligent Systems, IEEE [see also IEEE Intelligent Systems and Their Applications] **21** (2006), no. 6, 30–37.

17. D. Walton and C. Reed, *Diagramming, Argumentation Schemes and Critical Questions*, Anyone Who Has a View: Theoretical Contributions to the Study of Argumentation (2003).

18. D.N. Walton, *Argumentation Schemes for Presumptive Reasoning*, Lawrence Erlbaum Associates, 1996.

19. D.N. Walton and E.C.W. Krabbe, *Commitment in Dialogue: Basic Concepts of Interpersonal Reasoning*, State University of New York Press, New York, 1995.

20. J. Wanner, *Stable Foams and Sludge Bulking: The Largest Remaining Problems*, Journal of the Chartered Institution of Water and Environmental Management **12** (1998), no. 10, 368–374.

21. M. Wooldridge, *Introduction to Multiagent Systems*, John Wiley & Sons, Inc. New York, NY, USA, 2001.

Pancho Tolchinsky
Software Department
Technical University of Catalonia
Jordi Girona 1-3
Barcelona 08034
Spain
e-mail: `tolchinsky@lsi.upc.edu`

Montse Aulinas
Software Department
Technical University of Catalonia
Jordi Girona 1-3
Barcelona 08034
Spain
e-mail: `montseaulinas@gmail.com`

Ulises Cortés
Software Department
Technical University of Catalonia
Jordi Girona 1-3
Barcelona 08034
Spain
e-mail: `ia@lsi.upc.edu`

Manel Poch
Laboratory of Chemical and Environmental Engineering
University of Girona
Campus de Montilivi s/n
Girona 17071
Spain
e-mail: `manuel.poch@udg.edu`

Whitestein Series in Software Agent Technologies and Autonomic Computing, 61–90

Designing an Information System for the Preservation of the Insular Tropical Environment of Reunion Island

Integration of Databases, Knowledge Bases and Multi-Agent Systems by using Web Services

Noël Conruyt, Didier Sébastien, Rémy Courdier, Daniel David, Nicolas Sébastien and Tiana Ralambondrainy

Abstract. Decision-makers who wish to manage Insular Tropical Environments more efficiently need to narrow the gap between the production of scientific knowledge in universities, or other labs, and its pragmatic use by the general public and administrations. Today, one of the main challenges concerning the environment is the preservation of the biodiversity of ecosystems that suffer from urban and agricultural pressure. As we can only protect what we know, it is all the more important to share expert knowledge about habitats and species by using Internet in order to educate the public about their wealth and beauty. Based on Reunion Island, and taking into consideration an expected population growth of over 30% in the next twenty years, we are working to predict the human impact on this closed territory. To help tackle these two questions about biodiversity and land consumption, we have designed an Information System (IS) in the framework of the ETIC program. Our aim is to enhance insular tropical environment research in order to help the Reunion National Park to manage its protected territory. On the one hand, biodiversity research is handled statically, using knowledge bases and databases, to enhance Systematics and ecological university research. On the other hand, spatial planning concerns are treated dynamically, using multi-agent systems to simulate population densification movements. These software technologies have been implemented and integrated through a common architectural system in the ETIC program. They were conceived using Web Services that allow each module to communicate its functionalities and information with one another, as well as with external systems.

Keywords. Information System, Insular Tropical Environment, Biodiversity, Spatial planning, Multi-Agent Systems, Knowledge Bases, Databases, Web Services.

1. Introduction

Insular ecosystems are particularly rich, with remarkable endemism rates, but they are also extremely fragile and often highly deteriorated. In order to better protect biodiversity and natural spaces, expertise on these ecosystems needs to be propagated using information and communication technologies so that the most recent updated information may reach every-day people and activists. The more people know and understand their natural environments the more they will respect them and become emotionally attached to them. Policy makers, experts and members of civil society representing European, national, regional and local levels are aware of the necessity to preserve the patrimony of tropical islands [44]. The natural heritage of Reunion Island is rated as one of the official "hot spots" (Figure 1) in terms of world biodiversity [23].

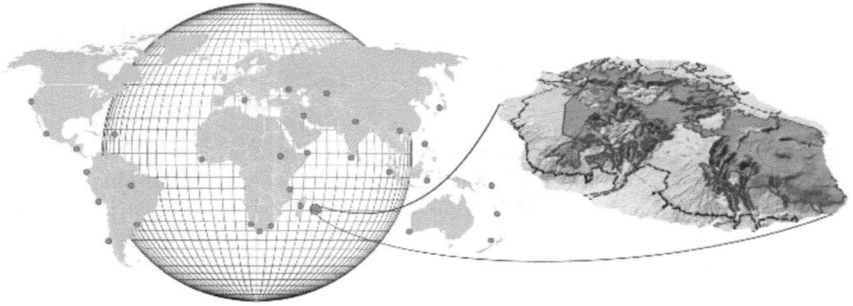

FIGURE 1. Reunion, a "hot spot" with its Natural Property (in dark grey).

The biological diversity of the islands in the South West Indian Ocean (Madagascar, Comoros, Mauritius, Reunion, the Scattered Islands) is still rich despite important anthropic pressure, which is increasing from year to year. The Reunion National Park was created in March 2007 in order to fix the limits of Natural Property (42% of the island's surface). Henceforth, authorities are preparing to apply for World Heritage Status in 2009 [31]. The Green Energy Revolution [43], a vision of Reunion Island in 2030, is another governmental project that is to deal with energy production and storage, high quality environmental habitats, and intelligent transportation and ecological tourism. The problem is that communication on these projects has thus far been unidirectional. The general public needs to be more involved in planning in order to optimize the population's participation towards the sustainable development of their island. There is a real need for investment in mid and long-term communication services. It is not sufficient to merely focus on saving energy and managing resources. For the future of ecology, the exchange of information is the key towards awareness of our shared natural heritage. The challenge for Reunion Island's inhabitants is thus to manage territory that is closed, just as the world itself (Figure 1).

There is a need for collective tools if we wish to manage common property [40]. The quality of information and the way it is delivered is of utmost importance in the shaping of public opinion. The university's role is thus to deliver qualitative data and expertise in order to help decision-makers make the right conservation choices, as well as to help inform and educate the general public. The University of Reunion has accumulated a large amount of qualitative observations, data, information and knowledge on ecosystems over the past forty years. This information may be found in laboratory checklists, collections, museums, literature, charts, maps, images, video files, audio files, and individual databases, yet is hardly exploited by anyone except the authors themselves. Moreover, in environmental studies, progress in research often depends on the information produced by social sciences, such as sociology, economics, and law. Yet, for various reasons (methodological, communication problems, intellectual property rights, etc.) the exchange of information between these different fields of study is still facing difficulties. New solutions are needed to treat global and complex problems in the information age.

The ETIC program was created in order to find solutions for research and knowledge enhancement of natural Insular Tropical Environments using Information and Communication Technologies [45]. It is based on collaborative methodology, stressing partnerships with researchers and associations who wish to share their skills in Systemics, Geomatics, Biomatics, or other domains of knowledge engineering and collective intelligence, by using telecommunication and computer science, with content producers, editorial boards, scientists, educators, decision-makers, enterprises and end-users.

This chapter will briefly present the insular tropical environment context of Reunion Island. We will explain the enhancement methodology we have set up, and the architecture used to design our IS for environmental preservation. We will illustrate this with an example and describe some software components and services that are part of our artificial intelligence research for the management of biodiversity and environmental sciences. All these components are presented from a methodological viewpoint in order to address questions concerning insular tropical environments. Finally, we will introduce the Web Services that we have developed to link applications and services, and discuss the next step for the IS in the fight to protect our natural heritage, by illustrating its usefulness for decision-makers in regards to the Reunion National Park [46].

2. The preservation of Reunion's insular tropical environment

Reunion is a 2,500 sq km island located in the South West Indian Ocean (Figure 1). In the coming years, this french overseas department will have to deal with numerous growth phenomena and with their related consequences. With regards to land development, Reunion is facing the challenge of accommodating an ever-greater population while, at the same time, preserving its agricultural soil and its exceptional landscapes and local species.

By 2030, the island population will increase from $775,000$ to $1,000,000$[47], which represents an impressive increase of over 30% in 20 years! This demographic trend leads to a number of problems, especially concerning housing. Taking into account the $225,000$ additional people, the need for housing creation is estimated at $150,000$. Thus, even if one puts forward a hypothesis of high density, the future of the island will result in an inevitable increase of the demand for urban land.

The rush towards urbanization is leading to greater anthropic pressure, which is increasing from year to year, on natural and agricultural spaces. This is an important problem for land and ecosystems management of the territory. Indeed, $50,000$ hectares are occupied by agriculture that, with its $15,000$ related jobs, has a powerful influence on the island's economy. Protecting the agricultural sector, including sugar cane plantations (the most important exploited resource), is therefore an additional priority; it is necessary to limit their consumption by urbanization and to prevent urban sprawl on the best farmland. Likewise, there are more than $100,000$ hectares of natural areas, harboring fragile species and rich biodiversity that need to be protected and controlled.

Henceforth, one of the main priorities for Reunion's government administrations (Regional and Department Councils, the Prefecture, the Reunion National Park) is to improve ecosystem management that is conditioned by ZNIEFF[48] inventories. The island's heritage, whether it be the land ecosystem (such as tropical forests) or the marine ecosystem (such as coral reefs) is altogether subject to growing economic pressure (urbanization and structural spatial projects, agricultural and industrial development) and to large-scale impact from tourism and leisure activities. The function of these natural areas is of the highest importance for the ecological equilibrium of the island as a whole.

Considering the situation from the perspective of sustainable development, urbanization policy concerning land management on Reunion Island needs to be as clear as possible. Many administrative documents treat urbanization at different levels on the island. Each one has its own importance because ecosystems management has to be considered both on the macro and micro levels. The SAR, Regional Land Development Plan, and SCOTs, Territorial Coherence Schemes, define large areas that have to be protected from urbanization according to global needs of biodiversity, while the PLUs, Local Urbanization Plans, deal with the localization of very specific areas, and species, to be protected from anthropic pressures in cities and districts.

In this context, territorial futurology, creating models and tools that will give us indications about the island and its evolution, is of capital importance. These models, which are part of the ETIC program, offer the possibility of testing the coherence of different management scenarios, and are intended for the use of policy-makers as tools to assist them during the decision-making process regarding choices that will effect our tomorrow, such as the elaboration of the SAR, SCOTs and PLUs, and their effective execution over the years.

3. Designing an Information System for environmental protection

ETIC is a publicly funded data and knowledge enhancement program, based on Reunion Island, whose goal is to develop innovative ideas and ICT solutions for the management of biodiversity, ecology and ecosystems research contents. As part of the natural environment, biodiversity has been defined as "the variety of life in all its forms, levels and interactions. It includes ecosystem, species and genetic diversity" [19]. The program was created in 2004 at the University of Reunion for research enhancement of Insular Tropical Environments, by using Artificial Intelligence techniques such as *Knowledge Engineering* for building expert systems, *Collective Intelligence* for building multiagent systems, and *Information and Communication* tools such as content management systems for sharing information. Indeed, the first step to protect our insular tropical environment is to better educate citizens about its richness because we can only protect what we know!

ETIC is based on several thematic projects and a collaborative methodology, stressing partnerships between researchers, educators, decision-makers, enterprises, associations and end-users who wish to share and communicate their environmental data and knowledge off and online. With the help of computer scientists, web designers, programmers and graphics experts, the common goal is to participate in the construction of an *Information System* (IS) for environmental management on the Internet. Contents include terrestrial and marine biodiversity descriptions about specimens, species geography, ecology, photography, taxonomy, and bibliography contextual information on Reunion Island.

3.1. Enhancement methodology

Our approach is intended to be interdisciplinary, entrepreneurial and constructivist [1]. It combines experimentation and theory in a network of exchanges so as to satisfy and anticipate the use of scientific contents by a variety of socio-economic players.

The ETIC program is structured according to enhancement projects, founded on the meeting of professionals (producers and editors), and the anticipation of needs of end-users in tropical insular environments. This participatory approach on the part of researchers is strongly motivated by the possibility of adding value and distributing their research results through product-services on the Web[1]. With the support of project leaders, the generalization of this approach to other content areas within the program has considerable potential.

Knowledge enhancement and diffusion has two forms within the IS:

- A collaborative local network site for researchers that brings together tools (software and services), functioning within a secured infrastructure (Intranet), in order to model interoperable data and knowledge. This *upstream enhancement* or e-research activity is an iterative and creative process for the development of application generators.

---

[1] Product-services are also called e-services in the digital age: they are internet based electronic services

- An Internet portal accessed by the general public, which is divided into two parts "sea and land" with the development of applications in thematic projects, *i.e.* instances of the tools mentioned above. This *downstream enhancement* site offers information relative to environmental questions and presents e-learning activities.

Upstream enhancement

The Intranet portal allows for the enhancement of scientific information (knowledge and data) within a common architecture while assuring their interoperability through Web services. For biodiversity monitoring, systematicians and biologists use IKBS (Iterative Knowledge Base System) as static personal knowledge management software on the micro level of specimens and taxa[2]. For spatial analysis and visualization, ecologists and geographers use ArcGIS (ESRI Geographic Information System) for data storage and management software on the meso level of populations and biotopes. For regional planning, specialists in Systemics and Geomatics use GEAMAS-NG (Generic Architecture for Multiagent Simulations - New Generation) as dynamic collective knowledge management software at the macro level of habitats and ecosystems. An information management scenario on natural forest biodiversity preservation will be explained in section 3.3.

For interoperability of marine and terrestrial information, the other objective of upstream enhancement is to represent metadata of applications in a structured and standardized way. The functionalities of different applications can be generalized in software and services by an inductive process that enriches them as common tools dedicated to more general tasks. The personal, ecological, photographic, geographic and terminological data and documents of biology experts are stored thanks to five dedicated database modules: a directory to authenticate users, a biodiversity module to monitor specimens and taxa, a multimedia module to share documents, a cartographic service to geo-reference data and a thesaurus to define specialized terms and illustrate them.

Downstream enhancement

The Internet portal presents thematic projects proposed and led by independent volunteers: Herbarium, Natural Risks, Littoral Information System (SIL), Medicinal and Aromatic Plants (MAP), Tropical Environment Management (GET), collective management of animal waste (BIOMAS), Coral Reef Monitoring (COREMO), Hydrogeology of the "Piton de La Fournaise", etc. These applications may be developed in collaboration with non-university partners such as the ARVAM [49] organization, or the VO [50] association.

For example, "Life in the coral reef" is a structuring project which proposes to unite the skills of content producers (University of Reunion laboratory researchers, educators from independent groups for nature conservation), content

---

[2]Taxa are the names of ranks in the scientific classification: Species, Genus, Family, Order, Class, ...

editors (IREMIA, Multimedia Centre) as well as professionals and students working within master's degree programs in Computer Science, Indian Ocean Communication, Tropical Environment Management Sciences, or Computer Graphics schools (ILOI [51]). The potential for information and communication calls for an analysis of the usages[3] of projected services on a co-design platform, uniting all at once the ETIC community of project leaders, researchers and users.

This constructivist approach gives new ideas for software development to computer scientists in return. For example, in order to manage coastal zones, we can offer access to a complete series of data concerning biodiversity, beach erosion, coral bleaching, protected areas, and drainage basins in a Littoral Information System (SIL). This general application is connected to ArcGIS that may be consulted or modified in the Intranet by biogeographers. As we also develop and enrich knowledge of experts in Systematics with knowledge bases such as "Corals of the Mascarene archipelago", which is done by using IKBS, the identification tool should communicate with SIL by offering access to species information on a map of reef biotopes. In the end, we are able to create behavior simulations with multiagent systems, for example on the interaction between corals and fish, or a chemical pollution intrusion in the lagoon, etc. The finality is that these tools become instruments in end-users' hands, *i.e.* really used in every day contexts [10].

3.2. Architecture of ETIC Information System

All of these marine and terrestrial thematic applications are at different levels of development: mock-up, prototype, product-service[4]. They can make use of transverse services such as the Directory for authenticating users. Some of them are founded on a database management system developed in PHP-MySQL but are not yet interoperable. They can also rely on software for knowledge base management (descriptive modeling with IKBS), geographic information gathering (spatial analysis with ArcGIS) or multiagent system behavior simulation (GEAMAS-NG).

The whole constitutes the IS for Tropical Insular Environment Management Support, of which the proposed functional architecture may be found in Figure 2. It is modeled as a SOA (Services Oriented Architecture) middleware with a hub of Web Software Services (WSS) and Web Component Services (WCS) on demand. This hub of Web Services is detailed further in section 6.

---

[3]In the context of a user-centered design research at the University of Reunion, a creative or co-design platform is a physical or virtual meeting and communication space for product-services to be developed with user input concerning expected needs being a key factor in the creative process. These e-services are developed in projects by a team of content producers (researchers), packaging editors (designers and programmers) and distributors of the final product (operators). The multimedia platform is similar to those which are met in the film and broadcasting industry (TV-Net style). Products and services are developed from a well-focused response to clear questions with specific tasks to be solved, with the help of end-users so that the new products most directly match the expected uses of the projected services [37].
[4]A product-service is an application or software tool that is really used in the domain area, whereas a prototype is something usable or occasionally used.

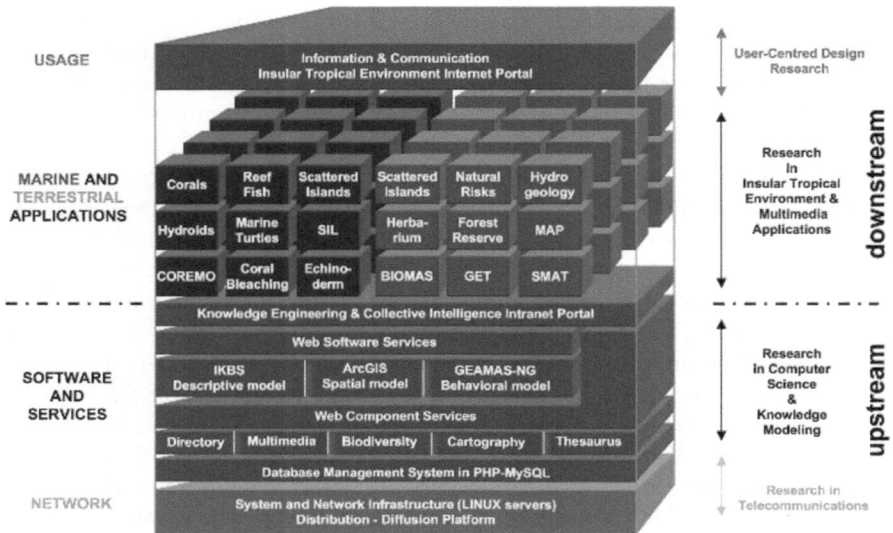

FIGURE 2. The functional architecture of ETIC Information System

3.3. Example

The ETIC program is structured according to projects for which the long-term objective is to make them interoperable in relation to environmental questions. For example, concerning the protected areas of the Reunion National Park, it is imperative to define the specific elements of biodiversity in the primary forest zones, the types of anthropic pressure on them (pests, urban scattering) and the expected evolution in 10 year's time. The first step is to be able to identify the plant specimens, at least their Genus names. The scenario in Figure 3 shows the tools that we are developing to help find answers to these questions.

Biodiversity information comes from expertise in Systematics and Ecology. In order to manage specimen and taxa information, we rely on knowledge bases and databases. These technologies are used in order to get all the descriptive and *static* details of biodiversity information. Different modules for collecting ecological, geographic, taxonomic, photographic and terminological data have been designed to constitute authored information sources. But to access this general information, one important entry is to know the name of the specimen under study. For this task, an iterative knowledge based system called IKBS has been built to help specialists in Systematics define descriptive models of a domain. For example, this could be a genus of orchids for which one may describe cases, and let people identify the name of a specimen through questions. These two steps form the static kernel of data and knowledge acquisition of our ETIC IS. It is complemented by

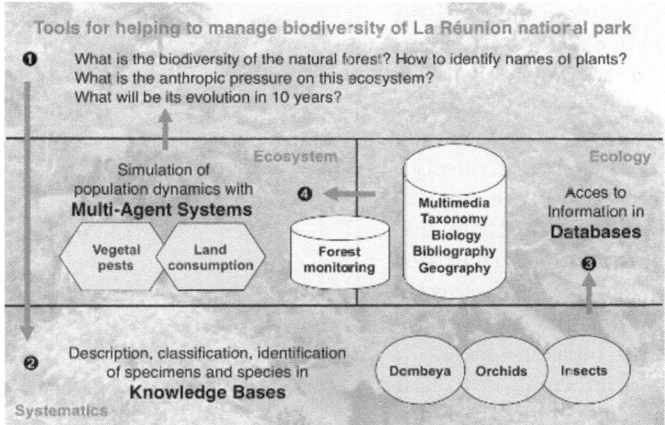

FIGURE 3. The information management service on natural forest biodiversity

a temporal follow-up of orchid species and spatial analyzes of the protected areas through a private Geographical Information System, ArcGIS [52].

When the wealth of a natural zone is known, decision-makers need other tools in order to manage these ecosystems - to analyze the human consumptions of the territory that comes from the urban and agricultural pressure on these habitats. The objective of SAR, SCOTs and PLUs is to explore scenarios for the evolution of the growing population. In our IS, these simulations are dealt with using a multiagent system called GEAMAS-NG that tests behavioral models of agglomerations, i.e. urban extensions. This simulation process constitutes the *dynamics* aspect of our method.

The next part of this chapter will present both static and dynamics aspects of environmental information management. After a brief overview of biodiversity descriptive knowledge and data management software and services, we will focus more precisely on modeling dynamics using multiagent technologies.

4. Modeling biodiversity static information with knowledge bases and databases

Systematics is the scientific discipline that deals with listing, describing, naming, classifying and identifying living organisms [22, 42]. In our research, we focus on populations of specimens between the taxa and organ levels of biodiversity research [20] (Figure 4).

The originality of our insular tropical biodiversity management method is that we concentrate on natural objects that are specimens in the field (living specimens), as well as specimens in museums (collection specimens). Experts in biology at universities have studied them intimately for years and are the only persons able to correctly identify species, which is an important step in the process towards offering access to more specialized information to non-experts. Researchers build their personal or tacit knowledge [29] by ob-

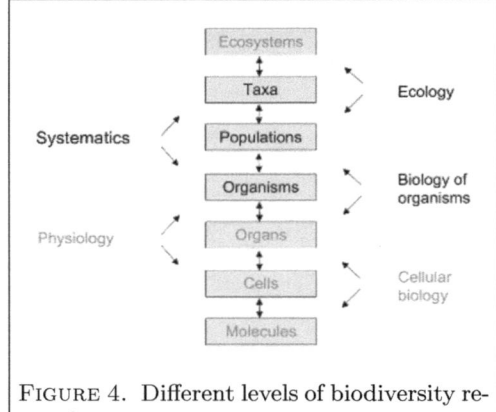

FIGURE 4. Different levels of biodiversity research

serving species in the field and in their laboratories under the microscope, then interpreting them with descriptions. These described objects form the basis for the development of their formal or explicit knowledge in monographs that constitutes their authority in their area of specialty.

The core of the ETIC IS platform is the integration of knowledge bases and databases about biodiversity knowledge and data by using Web Services.

4.1. Knowledge bases

In this part, we will briefly summarize the main functionalities of our knowledge engineering method. The complete methodology of knowledge management can be found in [9]. Knowledge base applications are instances of a Knowledge Based Management Tool called IKBS. This Iterative Knowledge Base System lets specialists define an Object-Attribute-Value morphological descriptive model of the domain knowledge (input), and describe cases (output) based on this ontology (Figure 5). The knowledge acquisition

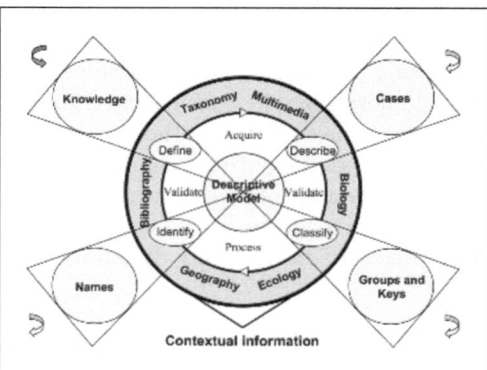

FIGURE 5. Biodiversity knowledge bases management system within the IS

phase can be repetitive because IKBS applies the scientific method of Popper [30] in biology (conjecture and test) with an iterative process of knowledge management (Figure 6):

1. observe and familiarize oneself;
2. represent observations, *i.e.* make a descriptive model and related cases;

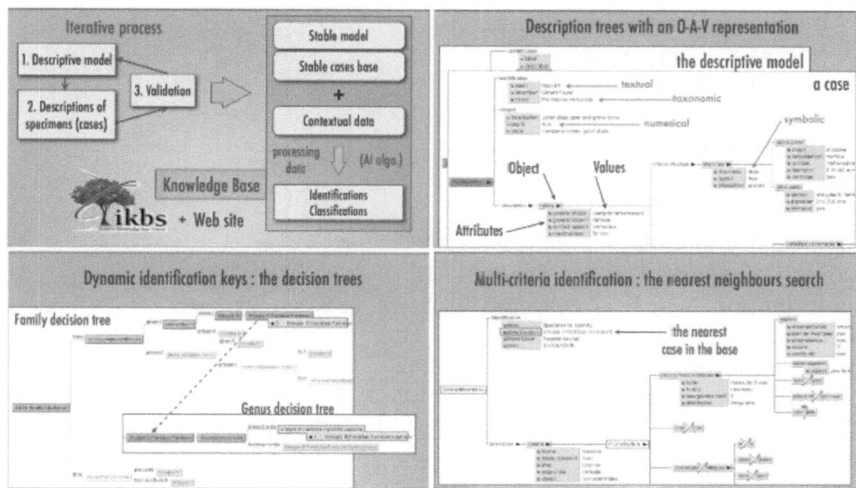

FIGURE 6. IKBS methodology of Systematics knowledge management

3. build hypotheses from pre-classified descriptions, *i.e.* generate identification keys (supervised classification);
4. test and use them with new observations, *i.e.* identify new specimens;
5. refine the initial knowledge (new characters, cases and classifications).

After an automatic classification process based on tree induction of pre-classified cases, end users are able to identify new descriptions with a questionnaire. End-users proceed by photo-interpretation of specimens to obtain a genus name, or by observing microscopic specimen elements under a binocular microscope to identify a species name [7].

4.2. Databases

The main database is a biodiversity module, *i.e.* a database of objects that stores, organizes and presents scientific data about field observations, collected specimens (samples) and taxa descriptions. Other database modules (*i.e.* directory, multimedia, thesaurus and cartography) complement this central module:

1. the directory gives access to the subjects, *i.e.* the individual and community researchers with their profiles in a card index;
2. the multimedia database manages all types of documents (photo, video, sound, etc.) that can be indexed to specimen objects and taxa;
3. the thesaurus will be an illustrated glossary that stores the meaning of Insular Tropical Environment vocabulary;
4. the cartography is a tool for georeferencing data on a map (made with the GoogleMaps API).

All of these modules are linked by Web Services so as to constitute a modular, interoperable and integrated biodiversity specimen and species database management system (Figure 7).

The data entry process in the biodiversity module is organized around the memorization of specimen information, which is collected in notebooks by biologists when they inventory biodiversity. It has been structured in five edition tasks (actions) that follow the daily work of monitoring specimens in the field:

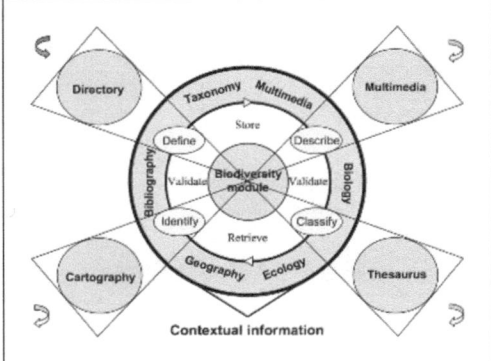

FIGURE 7. Biodiversity database management system within the IS

1. Origin of specimen, where was it found?
2. Short description, what was observed?
3. Taxon identification, what is it?
4. Status of specimen, *i.e.* sex, nature, state, fertility, development stage.
5. Label of specimen if it is to be put in collection.

FIGURE 8. ETIC process of biodiversity data management

In addition to this internal specimen information in the biodiversity module, the surrounding modules manage external contextual data such as the location of the specimen (space identification with the geolocation system), the identity of the subject (who is the observer in the Directory), the image or video of the specimen with associated metadata in the multimedia database (Figure 8).

5. Modeling ecosystem dynamics with agent technology

This section, introduces first the interest of using multiagent systems for ecosystem dynamics representation, then the behavioral model proposed and implanted in the GEAMAS-NG simulation toolkit is detailed. Finally, several examples of multiagent systems participating in the ETIC IS are presented.

5.1. Agent technology and complex systems

Complex natural and social systems are the subject of many studies aimed at understanding their characteristic dynamics and functioning. The frightening complexity of these systems constitutes a challenge for computer science modeling [35].

Understanding and Analyzing complex systems

The complexity of a system involves an important number of components, which are joined in such a way that it is difficult to separate them. This duality determines two dimensions of the complexity [3]:

- First, the distinction of several components imposes the obligation to find a way to reduce, to organize the system complexity, and to describe suitable components.
- Second, the connection between these components involves dynamic aspects of the system: they are linked with interactions that are dynamic relationships by means of a set of reciprocal actions among components. Interactions consist in exchanges of actions, from which the result of the system will emerge.

Therefore, a complex system is internally driven by interactions between its components whose result exceeds the contributions of individual components. In complex systems, individual components make decisions in accordance with various rules, on the basis of local, rather than global, information. Computational tools should provide intelligent capabilities to simultaneously integrate, in the same frame, a variety of information, and to perform a synthesis of events, which may follow. The boom of computer science in systems modeling during the past twenty years has greatly increased the understanding of complex systems by using virtual simulation [5].

The most notable difficulty of multiagent systems is to organize the dynamics of the ecosystem in question as a society of interconnected autonomous agents, that matches the complexity of the real world, and then to evaluate the results of the simulation. Figure 9 synthesizes some of the key activities by defining an iterative incremental cycle used to control the whole process of behavioral model development built with the help of relevant experts.

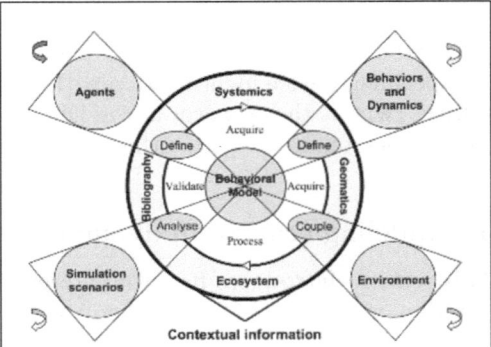

FIGURE 9. Multiagent modeling and simulation within the IS

This cycle describes four major activities:

1. Agent identification and definition of their attributes.
2. Defining behaviors that organize all the actions and influences the agents can undertake (with a methodic attention upon contextuality of the interacting elements).
3. Associating spatio-temporal environment to the agents.
4. Processing simulation scenarios on specific existing ecosystems on Reunion Island.

After each cycle, the simulation results are analyzed by relevant experts who validate the model. This leads to a new cycle, which aims at refining existing dynamics or at inserting a new dynamic in the model.

Emergence of phenomena

We have seen that what makes a complex system is the presence of interactions among the individual components of the system. But another property is the effect these relationships have on the behavior of both components and system. In complex systems, these relationships are at the root of emergent phenomena and constitute the key feature in almost all cases.

Emergence refers to the way the interactions among system components generate unexpected global system properties or behavior not present in any individual component taken separately. The treatment is centered upon the bottom-up approach, which offers a path of discovery towards possible solutions. While individual components of the real world are relatively simple in their behavior, when interacting, the collection of components turns out to be a richer structure, having a level of complexity greater than any one of its parts.

A good example of emergence is the organizational structure of an ant colony [16], achieving things that no individual ant could accomplish. All activities are carried out by individual ants acting in accordance with simple local rules and information. There is no master ant overseeing the entire colony and broadcasting instructions to individual ants. Interactions among ants give rise to patterns of global work allocation, which could not be predicted nor arise in any single ant.

Another example is that of shoals of fish that appear in lagoon simulations under particular conditions. These emerging shoals, which have their own behavior - collective behavior - exist in the simulation system instead of the hundred or thousand individual fish belonging to them. This provides significant interest for system comprehension, and offers the advantage of a drop in complexity during simulation.

One of the important characteristics leading to this kind of organization is that of the conditions under which a phenomenon emerges. In geophysical complex systems, this point of view has led to the concept of Self-Organized Criticality [2], to explain the "repeatability" of phenomena in nature. Such systems are driven by highly non-linear behavior; a small external perturbation could generate a large-scale phenomenon at a critical state of the system, but without predicting when it may appear. The system is therefore managed by a property that is unpredictable, but at the same time, the appearance of the phenomenon arises from instabilities, in which a small change in a component state can unbalance the whole system state, causing disasters. The critical state is thus seen as the trigger for emergent phenomena. This consideration gives importance to individual actions, which work toward the elaboration of the phenomenon, and therefore its organization.

One important result of this approach is to consider emergence (*i.e.* an emergent phenomenon) as a self-organized structure. This leads us to assimilate emergence in terms of self-organization. A computational model intended to develop simulation applications of complex systems should then propose an architecture in which emergent phenomena are dynamically created during simulation as they appear [18]. The result of the simulation is, on the one hand, interpreted in terms of quantified results, and on the other hand, assimilated as self-organization of new patterns which model emergent phenomena.

Dynamically creating structures is interesting because it allows the system to keep track of phenomena, while they are being adapted throughout the simulation. For instance, in natural phenomena, the affected part of the system can, in its turn, influence the other components; it plays a role in the system for future behavior. An example is the case of a vegetal pest spreading through a primary forest. This emergent phenomenon radically modifies the forest as it will never again return to its primary state. Our assumption is thus based on the fact that once it appears, an emergent phenomenon becomes intrinsic to the system, and its new characteristics can no longer be inferred in the same way as before.

Multiagent software engineering issues

Software engineering issues require one to produce a complete toolkit, a virtual laboratory, that can design a large scope of dynamic systems, study the informational structure of complex systems and provide generic interfaces to set and control the simulation.

As the system complexity cannot be globally expressed, the challenge is to find a computational model able to:

- Represent and distribute complexity in individual elements.
- Represent the system dynamics as local interactions between agents.
- Provide editing tools to build simulation scenarios.
- Provide mechanisms so that simulation results emerge by interpreting local interactions.
- Provide optimized mechanisms to support large-scale simulation, which implies throng interacting agents.

Propositions addressing these issues have been made through the GEAMAS simulation toolkit and its latest version, GEAMAS-New Generation. This version extends the capacity of the previous one to support a larger simulations scale (more than $100,000$ agents). GEAMAS-NG features:

- Dynamic-Oriented Modeling [26], that models real systems by representing their intertwined dynamics through a set of monothematic dynamic submodels.
- Temporality time management model [25] [38], that eases agent description and optimizes execution.
- Configuration tools [27], that provide scenario description languages and original map-based initialization.
- Emergence manipulation mechanisms [15], that enable reification of emergent phenomena that occur during ecosystems simulation.
- Parallel and distributed simulation execution [38], that supports large-scale simulation execution by dynamically distributing agents on the execution infrastructure.
- Advanced observation of simulation results [32], that provide pertinent visualization tools specific to agent concepts such as agent interactions and conversations analysis.

Considering that the aim of this section is to focus on modeling ecosystems dynamics, let us now elucidate the first item mentioned above - Dynamic-Oriented Modeling.

5.2. Dynamic-Oriented Modeling

Modeling a complex system into an agent-oriented simulation model is not an easy task as experts must define the agents that interact with each other in an environment[5]. Indeed, agents usually participate in several dynamics thus intertwining them. This makes modeling more complex. Dynamic-Oriented Modeling aims at modeling agents by identifying their interactions with the system dynamics.

---

[5] *environment* as to be considered, in this section, on its multiagent definition, An *environment* provides the conditions under which an agent exists. Different environment types exist. *Physical environments* provide those principles and processes that govern and support a population of agent. *Communication environments* provide those principles, processes and structures that enable an infrastructure for agents to convey information [24]

Modeling dynamics in a multiagent system

In this section, a dynamic is defined as an association of a set of activities that participate in the specification of a major characteristic concerning the study of a phenomenon.

To model the system's evolution, it is necessary to let some dynamics show through the behavior of the agents and the properties of the environment. Therefore, the greater the number of dynamics to consider, the more complex the design of an agent will be. We have described a modeling approach called dynamic-oriented modeling which puts dynamics at the center of the modeling process. This approach is a means to circumvent complexity.

The state of an agent is composed of all its attributes, and its behavior organizes all the actions it may undertake. These two sets can be divided according to the dynamic to which their elements are referred, and some *modify/influence* relationships can be established between the subsets obtained. Figure 10 illustrates a partition composed of 3 dynamics: A, B and C.

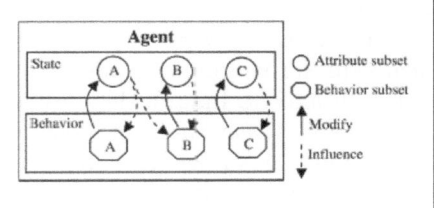

FIGURE 10. Agent modeling according to the different dynamics at stake

In the same way as the subset of attributes A shown in Figure 10, some of the attributes, which are linked to a dynamic often have an impact on the behavior in connection with another dynamic. These particular attributes are the prevailing attributes of the dynamic to which they refer.

Once this characterization obtained, the model can be divided into a number of sub-models equal to the number of dynamics to manage. Each sub-model is exclusively associated to one of the dynamics, and is called a Mono-Dynamic Model (MDM). The agent is instantiated in every MDM that models a dynamic in which the agent participates. Yet, in each MDM its state and its behavior are reduced exclusively to the subsets that deal with the dynamic of the MDM (Figure 11).

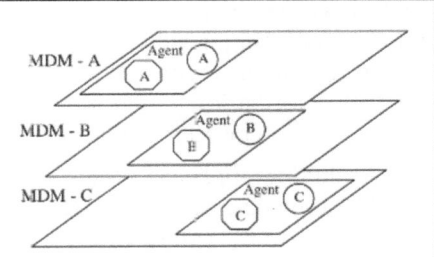

FIGURE 11. Splitting of an agent in a dynamic-oriented modeling

Coupling dynamics in a multiagent system

Multiagent simulation, in the ETIC Information System, is used to address issues concerning the management of ecosystems that have spatial aspects. In this context of spatial MAS [33], agents are associated with a part of the space in the environment.

This characteristic enables one to use the environment as a coupling element between the different MDM. To do so, it is necessary to extract the prevailing attributes of the MDM from the agent's subset of attributes, and affect them as properties of the surface associated with the agent. In this way, the information contained in prevailing attributes is available in the environment, and accessible by agents situated in other MDM (Figure 12).

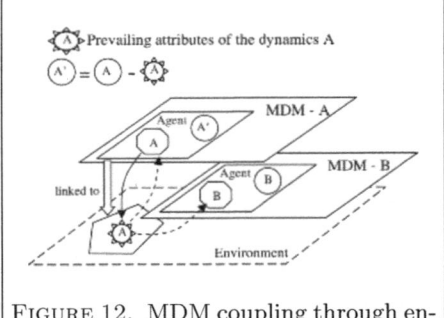

FIGURE 12. MDM coupling through environment

The *modify / influence* interactions between agents and prevailing attributes located in the environment are completely natural in the exchanges between agents and environment. Indeed, they result from the *actions / perceptions* that an agent can make in its environment [35].

A dynamic-oriented simulation thus leads to a layer-structured agent model. Each layer, called MDM, models one particular dynamic exclusively. Layers do not directly interact with each other, but use the environment as a place to share information. The results of the processing of a dynamic are written in the environment by the agents, located in the MDM representing the dynamic. These results are then perceived in the environment by agents of other MDM and can possibly be taken into consideration to determine actions they must undertake to process their own dynamic. In other words, this approach enables the organization of information in independent modeling layers. As each layer is independent and represents a particular aspect of the real system, it is possible to configure a layer by interconnecting it with an information repository (see section 5.3).

In this manner, the environment constitutes a flexible point where MDM can exchange information, while warranting independence between the MDM. In this context, the integration of one MDM does not affect other MDM.

A similar application can be conceived by replacing one of the MDM with a software layer, by associating the necessary calculations of the evolution of prevailing attributes with the dynamic that was previously managed by the removed MDM (Figure 13). This software layer can be, for example, a cellular automata, a virtual map interfaced with a GIS, or a stochastic grid of values, for example.

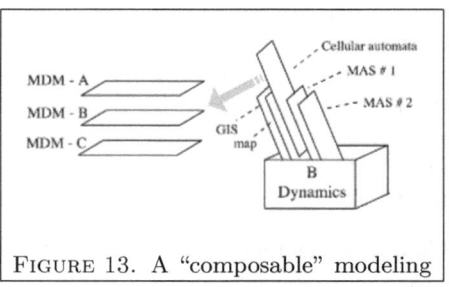

FIGURE 13. A "composable" modeling

5.3. Integrating heterogenous information layers

This dynamic-oriented modeling is a key feature for the simulation of biodiversity on the scale of a territory. Every dynamic of the system can be represented as

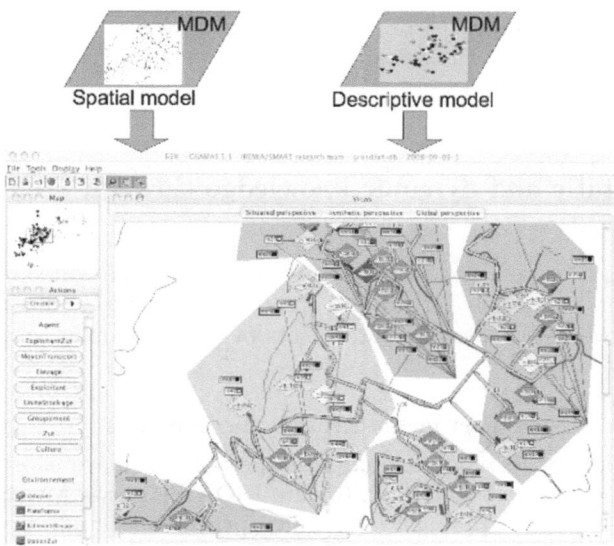

FIGURE 14. Biomas imports data from databases and GIS, and exports simulation results as maps in GIS

a Mono-Dynamic Model, for example, urban space, agricultural space, or natural space. Dynamics interact with each other through the environment. The multiagent simulation can then mix these dynamics to produce a global simulation. The results help policy-makers in their decisions concerning the ecosystem preservation. In collaboration with the CIRAD agricultural research center[6], we have developed two multiagent simulation applications based on this dynamic-oriented modeling.

The first one is the BIOMAS application [12], based on the GEAMAS simulation toolkit [21]. BIOMAS simulates the effect of agricultural dynamics on ecosystems. It simulates collective organic matter fluxes transferred amongst a set of farms located within a territory. This model is initialized by data taken from a descriptive model with detailed information about crops, farms and agricultural stakeholders. Moreover, the spatial environment of the agricultural stakeholders is also coming from an external spatial model (Figure 14).

The second one is the SMAT application [14], built according to the dynamic-oriented modeling introduced in GEAMAS-NG. It is composed of a set of mono-dynamic models, mainly urbanization models, natural models and agricultural models. In order to do this, the multiagent simulation model is initialized using static data about the ecosystem and the environment (see Figure 18 in section 6.4). These data are contained in external databases and in geographical information systems.

---

[6]French Agricultural Research Center for International Development [53]

With these two models, we have successfully run large-scale simulations on the scale of a territory for the dynamic evolution of a whole ecosystem, based on static data from the ETIC IS.

6. Binding static and dynamic knowledge through Web services layers

One of the major originalities of the ETIC IS is that the software, components and applications, previously introduced through web services in a coherent architecture, that covers all the modeling aspects of insular tropical environment, are interconnected.

6.1. Web services as support for integration and openness

In order to make our IS modular and interoperable, we have chosen to design it according to Services Oriented Architecture (SOA). As it is a web platform, we chose the Web Services approach [13], which provides a number of features and benefits.

Among the technologies used to implement Web Services, we chose the WS* technology whose specifications are based on SOAP and WSDL standards:

- SOAP (Simple Object Access Protocol) is used to exchange messages. It is a RPC (Remote Procedure Call) protocol built on object-oriented XML.
- WSDL (Web Service Description Language) is used to describe: Web Services, their operations, messages used, the types of used data, and the used protocols. The WSDL describes a public interface that provides access to Web Services. This description is written in XML and indicates "how to use the service".

6.2. Advantages of using Web Services in ETIC IS

Using Web Services structured the way we developed ETIC IS:

- Firstly, we have chosen this technology because it allows other IS to use our modules independently [28]. In this way, information stored in the ETIC IS is fully open and exploitable by other institutions [41].
- Secondly, it has modified the way we imagine connections in the IS itself. It has let us develop each module in a heterogeneous way, using the appropriate technology (PHP/JAVA/Flash Action Script). For example, the ETIC Directory is developed in PHP whereas the first version of our Multimedia Database [36] was relying on JAVA technology. Although we used different programming languages, these two modules were able to exchange secured information, thanks to Web Services.
- Thirdly, from a technical point of view, Web Services impose rigor that allows us to update our systems and services without losing the compatibility with older versions. Each new version of a service is indexed with a number, and

Web Services	Provider							
(Client)	Directory	Multimedia	Biodiversity	Cartography	Thesaurus	IKBS	ArcGIS	GEAMAS-NG
Directory	▨	✓	✗	✓	✗	✗	✗	✗
Multimedia	✓	▨	✗	✓	✓	✗	✗	✗
Biodiversity	✓	✓	▨	✓	✓	✓	✗	✓
Cartography	✓	✗	✗	▨	✗	✗	✓	✓
Thesaurus	✓	✗	✗	✓	▨	✗	✗	✗
IKBS	✓	✓	✓	✓	✓	▨	✓	✗
ArcGIS	✓	✓	✓	✓	✓	✗	▨	✓
GEAMAS-NG	✓	✓	✓	✓	✓	✓	✓	▨

FIGURE 15. Connections between ETIC modules

the WSDL provides a way to implement the new client, but it does not impact on the older versions that remain fully functional.

- Fourthly, from a management point of view, using Web Services was a good choice. The ETIC development team has often changed. Because of this high degree of turnover, it was difficult to transmit the key points of our development from the previous engineer to his successor. Thanks to the Web Services (and particularly the WSDL declaration), it was not necessary for the new team to fully understand a service source code to use it as a client. It saved a lot of development time.

Using Web Services was of course important to facilitate communication with our IS, but it also helped us to improve its inner structure.

6.3. What are the ETIC services?

Because each module and software is dedicated to accomplish a precise task, each one provides some specific functionalities and dedicated Web Services. In this way, any thematic application or other IS can reach ours and decide to use one or the whole of ETIC services for its own purpose. But contrary to data consultation that is completely open, these clients need to be referenced in the ETIC directory if they want to add information in the ETIC IS. This security allows us to ensure the traceability of information injected in our system.

The implemented connections between ETIC modules are shown in Figure 15. Of the five modules (Directory, Multimedia, Cartography, Biodiversity, Thesaurus) and three software (IKBS, ArcGIS, GEAMAS-NG) deployed now, all use Web Services as the client, and most of them provide services (Figure 16).

Work is still in progress. For instance, the current version of the Cartographic Browser does not implement a connection to the ETIC directory. In fact, it cannot be used as a stand-alone module yet. Other modules that connect to the Cartographic Browser must process client identification independently. Moreover, none of the functionalities that are supported by the modules have a Web Service offering the possibility of distant interrogation of it yet (e.g., the remote use of

		Tasks provided by Web Service
Components	Directory	• Find a user by his name • Get the list of members' identifiers (id) • Get a minimum of information on each member of the directory • Get a minimum of information on a specified member • Get all information about a specified member • Authenticate a member with its login/password (secured service)
	Documentary Multimedia Database (DMD)	• Return two links (documents' thumbnail and mini-thumbnail) corresponding to the input array of documents' ids • Return all information (metadata, download link) corresponding to the input array of documents' ids • Return two links (one to the information card on the DMD and the other to the file) corresponding to the input array of documents' ids • Allow the distant upload of a document on the DMD • Find a document on the DMD (Quick Search)
	Cartographic Browser	• Verify if a map exists for a specified id • Get all information about specified map • Find a map from its description • Delete maps from the given list
	Thesaurus	• Verify if a term exists and returns its id • Return an array of term's information corresponding to the requested ids • Return all information about one term owned by a specific user
Software	IKBS	• Build a decision tree and return the adequate question to optimize specimen's identification
	ArcGIS	• Return a processed map (analyze) • Convert Map's format
	GEAMAS-NG	• Return a processed map (simulation)

FIGURE 16. Functional Web Services in ETIC IS

the multi-criteria search of the MDB, see Figure 16). Nevertheless, most of the principal services of this module have already been developed.

Independent of the fact that we have to develop new Web Services for each module, we also have to increase the integration of the existing services between modules. This integration must be the result of focus groups realized with biologists in order to provide them the necessary information when they need it. This fine-tuning is a Web Design concern because, for end-users, Web Services must be transparent.

6.4. Two layers of services adapted to two kinds of needs

The ETIC IS relies on two different layers of Web services (Figure 17).

The first one, Web Component Services (WCS), is dedicated to provide convenient remote access to the modules. These modules may be considered as autonomous Web applications for thematicians, but also as toolboxes for developers. By using this service layer, they can focus on the front-end graphical interface of their own application without losing time in developing low-level processes. The second Web services layer, Web Software Services (WSS), offers access to the Knowledge Engineering & Collective Intelligence (KECI) portal. This layer analyzes the nature of the request and redirects it to the appropriate application. It fulfills advanced functionalities, like those described in the knowledge production layer. In this frame, the professional software suite "ArcGIS" introduces complex generic processes in order to add geospatial analysis to our platform.

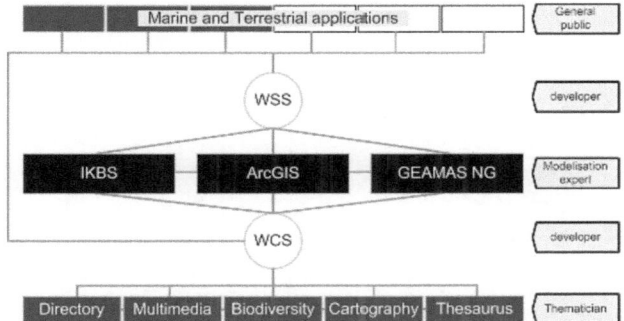

FIGURE 17. Global transactions between IS components.

If thematicians are able to feed and use modules on their own, they still need modeling experts' help to assist them, in order to use the knowledge production software and stimulate the emergence of new knowledge from data. We thus notice the need to connect modules (static knowledge) to the software (dynamic knowledge). From a technical viewpoint, Web service layers completely fulfill this need. But from a "human operator" viewpoint, the best results can only be obtained with teams of experts who master all three dimensions: Multiagent Systems, knowledge bases, and geospatial analysis. Interconnections between components and software, and between different software modules, are currently in an experimental stage, but already promise great results:

- GEAMAS-NG's agents can be initialized by maps derived from either the cartographic module or ArcMap, in order to set up interconnection configurations (Figure 18).
- IKBS's knowledge bases can help GEAMAS-NG during specific simulations. During a simulation, if agents representing spaces should be generated, IKBS is able to determine the best consistency for the emergent taxa.
- ArcGIS's algorithms can process spatial data from simulation results generated by GEAMAS-NG, and store maps for future use by policy-makers.

7. Results and discussion

Figure 19 shows the links between applications from downstream enhancement, on the one hand, and, on the other, software and services from the upstream enhancement of our IS. In the four-year period of the ETIC program between 2004 and 2007, we built fifteen applications, *i.e.* two product-services, corals [54] and hydroids [55], nine prototypes and four mock-ups [6]. The software IKBS and GEAMAS, which generate static and dynamic knowledge-based applications (Corals, Marine Turtles, BIOMAS) were already operational before the beginning of the program, but they have been improved thanks to user input from focus groups and questionnaires. To give an example, GEAMAS-NG was able to help

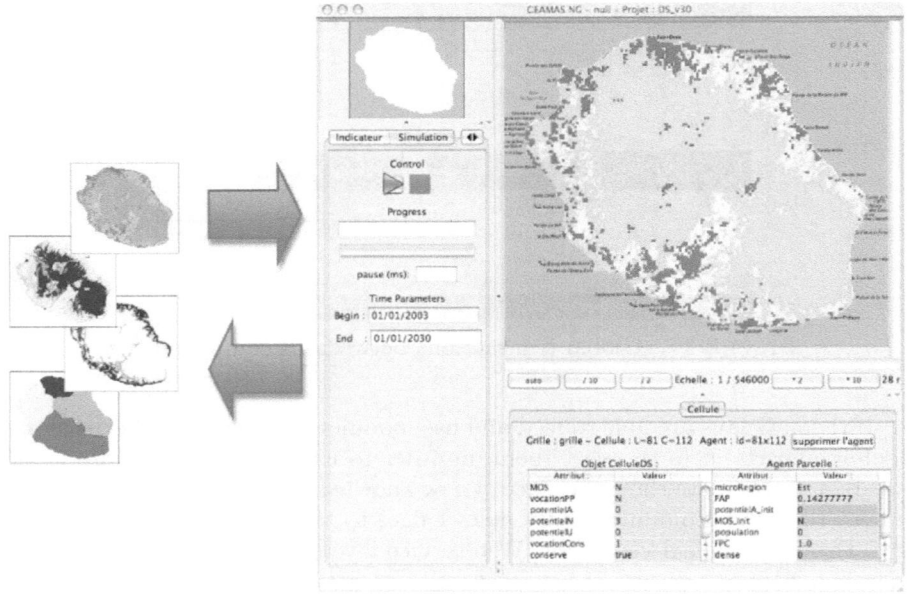

FIGURE 18. GEAMAS-NG is initialized and produces maps.

experts in designing and exploiting simulation models (see section 5.1) by providing concepts and user-friendly interfaces [32].

The first component service to arrive was the Directory in 2006 because the first demand was naturally and logically to promote the researchers themselves, and give them editing rights.

In 2007, we implemented the biodiversity and multimedia transverse modules that showed the desire of researchers to share their metadata. The first cartography module called FLIMBER [56] was made with Flash technology on ArcIMS, the Internet viewer of ArcGIS, but due to a lack of interactivity and map cover, it was replaced by a simple geolocation service made with the GoogleMaps API. This module was directly connected to the multimedia document manager (see Figure 8 of section 4.2).

The objective of the IREMIA team is to integrate the different data and knowledge modules as a hub of Web Software and Component Services to satisfy their interoperability and also their openness to other biodiversity and landscape information systems. Indeed, interoperability and integration of databases are the innovative challenge of international initiatives such as GBIF [57] on biodiversity, EDIT [58] on taxonomy, INSPIRE [59] on cartography. These information management projects are complemented with international projects on education and pedagogy through identification services such as KeyToNature [60].

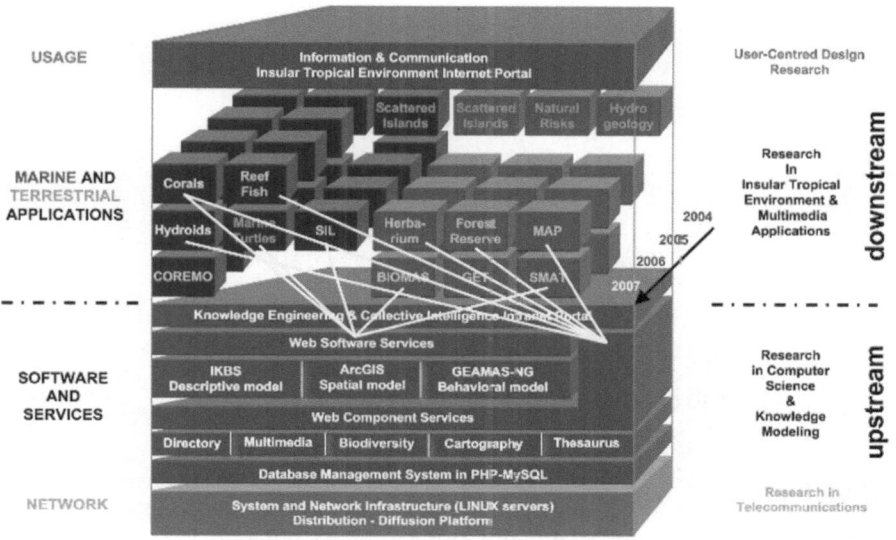

FIGURE 19. Instantiation of ETIC architecture in 2007

However, all these international projects are not sufficient if our goal is to create an IS that answers political, management and pedagogical questions on a regional and local level - one that becomes an Information Service (not only a System) for citizens of Reunion Island. The design of the Insular Tropical Environment Information System is based on a collaborative methodology between marine and terrestrial research and education producers, computer science and graphic design editors, as well as with enterprises and end-users. All these persons must meet together on a co-design platform, share a common vision, work together with some pleasure and constitute a community of practice. The Intranet portal made with SPIP was a good experience that prefigured Wikis as a virtual exchange space. Nevertheless, the success of each project depends on the motivation of the leader that has to spend time devoted to the animation and communication process of the future product-service. This task is the most challenging issue of knowledge management 2.0 in communities of practice, because if the first objective of managing people is resolved, the documents will follow more easily. The advent of web 2.0 tools in 2006 calls for the use of social networks to do intranet collaborative research work (CSCW or e-research), and to promote Internet collaborative public learning on a large scale (CSCL or e-learning) [11]. This orientation will be the next ETIC strategy for building the community of e-researchers and e-learners that define e-services together for insular tropical environment management and education (Figure 20).

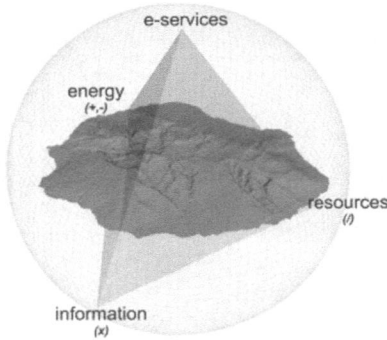

FIGURE 20. Philosophy of next ETIC program

8. Conclusion

Managing a commons is a very difficult and complex challenge, because the protection of our natural patrimony is the result of people's awareness and faith in their future. It is also the inherited responsibility of a society that reaches adulthood. The creation of the Reunion National Park in 2007 is a first step in this direction, because it was necessary to stop the uncontrolled development of anarchic and galloping urbanization. In order to truly preserve insular tropical environments, the decision help tools we have designed in our ETIC program will not be sufficient. Data and Knowledge management are preferably centered on users rather than on documents. This idea is derived from the organization 2.0 initiatives [34], also called the next enterprise that emphasizes trust, e-services and social networks. In fact, Information *Systems* must be replaced by Information *Services* in order to innovate - mostly by anticipating usages, and not only technologies. This is why we will name the next ETIC program Ne×Tic (New insular tropical environment and ICT), which will focus on the use of prototypes.

The ethics of our Ne×Tic program is to share services that combine the three dimensions of a sustainable development process: environmental energy, economic resources and social information (Figure 20). Environmental energy is bipolar (+,-): it determines the movement of things in one way or another. For the future, people in our insular tropical environment have to decide either to go in a material direction (*i.e.*, industrial, rational, economical and individualist) or to engage themselves in an immaterial direction (*i.e.*, post-industrial, symbolic, social and collective). The first choice is focused on consumption; the second one exemplifies creativity and openness. Indeed, when we share a material good, it divides itself, whereas when we share an immaterial good, it multiplies itself [39]. The problem is that resources and energy are becoming rare on our small island. We may say that Reunion represents a laboratory of the finite world with its natural and cultural diversity and beauty, but also its social difficulties. The equilibrium of a sustainable development model will not be possible without taking into account the ICT dimension, which is necessary to make connections between people in a social network. Information must be shared and discussed by citizens for them to participate in their future.

Acknowledgments

The ETIC program has been financed by the DOCUP-FEDER 2002-2006, A9-04 ICT measure from the French National Government, the Regional Council of Reunion and the European Union. We thank Mr. Michael Manoff for his help in translating this paper.

References

[1] E. K. Ackermann, *Constructing Knowledge and Transforming the World, A Learning Zone of One's Own* M. Tokoro and L. Steels (Eds.), IOS Press, 2004.

[2] P. Bak, C. Tang, and K. Wiesenfeld, *Self-Organized criticality: An explanation of 1/f Noise* Physical review Letter **59-4** (1987), 381–384.

[3] T.R.J. Bossomaier and D.G.Green, *Complex systems*, Cambridge University Press, Amsterdam, 1996.

[4] T. Cadet, *La végétation de l'île de La Réunion,* Thèse es sciences, Université de la Réunion, 1980.

[5] J.S. Carson, *Introduction to Modeling and Simulation* Winter Simulation Conference 2005, 2005.

[6] N. Conruyt, Activity report of ETIC program (in French, http://etic.univ-reunion.fr), IREMIA, University of Reunion, 2008.

[7] N. Conruyt, D. Grosser, Y. Geynet, *Enhancing Insular Tropical Environment contents through ICT and a Co-design methodology: the case of the ETIC program* 5th ECAI Workshop on Binding Environmental Sciences and Artificial Intelligence (BESAI'06), Riva del Garda, Italy, 2006.

[8] N. Conruyt, P. Conruyt, O. Sébastien, *A methodology for Designing E-services from a Co-Design Platform* 6th International Conference on Knowledge Management (IKNOW'06), Graz, Austria, 2006.

[9] N. Conruyt, D. Grosser, *Knowledge management in environmental sciences with IKBS: application to Systematics of Corals of the Mascarene Archipelago*, Selected Contributions in Data Analysis and Classification, Series: Studies in Classification, Data Analysis, and Knowledge Organization, 2007.

[10] N. Conruyt, O. Sebastien, D. Grosser, Methodology for defining e-services for knowledge management using a Co-Design Platform: example in the domain of instrumental e-learning, Common Innovation in e-learning, Machine Learning and Humanoid, 6th International Conferences on Human System Learning, ICHSL.6, Edited by: Sad Tazi & Khaldoun Zreik, pp. 117-124, ISBN 978-2-909285-46-4, IEEE, Europia (Eds), Paris, 2008.

[11] N. Conruyt, O. Sebastien, D.Sebastien, S. Calderoni, *How to Define E-Services From a Co-Design Platform in the New Web Philosophy: Application To Enhancement of Insular Tropical Environment and to Instrumental E-Learning*, Keynote presentation, 3rd International Conference on Open and Online Learning, ICOOL'2007, Penang, Malaysia, 2007.

[12] R. Courdier, F. Guerrin, F. Andriamasinoro, J.-M. Paillat, *Agent-based simulation of complex systems: application to collective management of animal wastes* Journal of Artificial Societies and Social Simulation **5-3** (2002).

[13] F. Curbera, M. Mukhi, S. Weerawarana, *On the Emergence of a Web Services Component Model* 6th International Workshop on Component Oriented Programming (WCOP'01), 2001.

[14] D. David, D. Payet, A. Botta, G. Lajoie, S. Manglou, R. Courdier, *Un couplage de dynamiques comportementales : Le modèle DS pour l'aménagement du territoire* JFSMA07, Carcasonne, France, 2007.

[15] D. David, R. Courdier, *Emergence as Metaknowledge: Refining Simulation Models through Emergence Reification* ESM08, Le Havre, France, 2008.

[16] A. Drogoul, J. Ferber, *multiagent Simulation as a tool for Modeling Societies: Application to Social Differentiation in Ant colonies* Artificial Social Systems **830** (1994), 3–23.

[17] G. Faure, *Recherche sur les peuplements de scléractiniaires des récifs coralliens des Mascareignes*, Thèse es sciences, Université Aix-Marseille II, 1982.

[18] J. Ferber, *Reactive Distributed Artificial Intelligence: Principles and Applications* Foundations of Distributed Artificial Intelligence, North-Holland, 1994.

[19] M.L. Hunter, *Jr. Fundamentals of conservation biology* Blackwell Science, Cambridge, Massachusetts, 1996.

[20] J. Lebbe, *Systématique et informatique* Systématique et biodiversité **13** (1995), 71–79.

[21] P. Marcenac, R. Courdier, S. Calderoni, J.-C. Soulie, *Towards an Emergence Machine for Complex Systems Simulations* Lecture Notes in Computer Science **1416** (1998), 785–794.

[22] L. Matile, P. Tassy and D. Gouget, *Introduction la systématique zoologique*, Biosystema 1, Paris, 1987.

[23] N. Myers, R. A. Mittermeier, C. G. Mittermeier, G. A. B. Fonseca, and J. Kent, *Biodiversity Hotspots for Conservation Priorities* Nature **403** (2000), 853–858.

[24] J. J. Odell, H. Van Dyke Parunak, M. Fleischer and S. Brueckner, *Modeling Agents and their Environment* Agent-Oriented Software Engineering III **2585** (2003).

[25] D. Payet, R. Courdier, T. Ralambondrainy, N. Sébastien, *Le modèle à Temporalité : pour un équilibre entre adéquation et optimisation du temps dans les simulations agents*, JFSMA06, Annecy, France, 2006.

[26] D. Payet, R. Courdier, N. Sébastien, T. Ralambondrainy, *Environment as support for simplification, reuse and integration of processes in spatial MAS*, IRI 2006 (Information Reuse and Integration), Hawaii, USA, 2006.

[27] D. Payet, D. David, N. Sébastien, *XELOC: eXtensible Editing Language Of Configuration - To facilitate complex systems configuration edition and reuse*, EA2525 LIM/IREMIA Technical Report, 2007.

[28] P.F. Pires, M.R.F. Benevides, M. Mattoso, *Building Reliable Web Services Compositions* Lecture Note in Computer Science **2593** (2002), 59–72.

[29] M. Polanyi, *Personal Knowledge: Towards a Post-Critical Philosophy*, Chicago: The University of Chicago Press, 1962.

[30] K.R. Popper, *La logique de la découverte scientifique*, Payot (Eds.) Press, 1973.

[31] PNRun, Pitons, Cirques and Remparts of Reunion Island: Aiming at World Heritage Status (http://www.reunion-parcnational.fr/spip.php?article45), Reunion National Park, 2008.

[32] T. Ralambondrainy, J.M. Médoc, R. Courdier, F. Guerrin, *Tools to visualise the structure of multiagent conversations at various levels of analysis* Modsim'07 International congress on modelling and simulation: Integrated systems for sustainability, Christchurch, New-Zealand, 2007.

[33] A. Rodrigues, J. Raper, *Defining spatial agents* Spatial Multimedia and Virtual Reality Research Monograph (1997).

[34] M. Roulleaux Dugage, Organisation 2.0. Le knowledge management nouvelle génération, Eyrolles (Eds), 2008.

[35] S. Russel, P. Norvig *Artificial intelligence: a modern approach* Prentice Hall, 2003.

[36] D. Sébastien, N. Conruyt, *Online multimedia database for communities of practice in Biology: a real use challenge*, 3rd International Conference on Internet and Web Applications and Services, 2008.

[37] O. Sébastien, N. Conruyt, D. Grosser, *Defining e-services using a co-design platform: example in the domain of instrumental e-learning* Journal of Interactive Technology and Smart Education **5-3** (2008), 144–156.

[38] N. Sébastien, R. Courdier, D. Hoareau, M.P. Huget, *Analysis of temporal dependencies of perceptions and influences ofr the distributed execution of agent-oriented simulations* ESM08, Le Havre, France, 2008.

[39] S. Soudoplatoff, Réseaux technologiques, réseaux humains, keynote given at Sophia-Antipolis Foundation (http://www.sophia-antipolis.org/fsa/animation/petits-dejeuners/2007/15-06-2007/internet.htm), 15th of June 2007.

[40] L. Steels, *Community memories for sustainable Societies* Sony Computer Science Lab, Paris, 2008.

[41] J. Yang, M.P. Papazoglou, *Web Component: A Substrate for Web Service Reuse and Composition*,14th International Conference on Advanced Information Systems Engineering (CAISE'02), Toronto, Canada, 2002.

[42] J. E. Winston, *Describing Species: Practical Taxonomic Procedure for Biologists.* New York: Columbia University Press, 1999.

[43] http://www.gerri.fr/, visited on 09/29/2008.

[44] http://www.reunion2008.eu/pages/en/en-home.html, visited on 09/29/2008.

[45] http://etic.univ-reunion.fr, visited on 09/29/2008.

[46] http://www.reunion-parcnational.fr/, visited on 09/29/2008.

[47] http://www.insee.fr/fr/regions/reunion/, visited on 09/29/2008.

[48] http://inpn.mnhn.fr/inpn/en/biodiv/znieff/index.htm, visited on 09/29/2008.

[49] http://www.arvam.com/, visited on 09/29/2008.

[50] http://vieoceane.free.fr/, visited on 09/29/2008.

[51] http://www.iloi.net/, visited on 09/29/2008.

[52] http://www.esri.com/software/arcgis/, visited on 09/29/2008.

[53] http://www.cirad.fr/en/, last visited on 09/29/2008.

[54] http://coraux.univ-reunion.fr/, visited on 09/29/2008.

[55] http://etic.univ-reunion.fr/hydroids/, visited on 09/29/2008.

[56] http://etic.univ-reunion.fr/flimber/, visited on 09/29/2008.

[57] http://www.gbif.org/, visited on 09/29/2008.

[58] http://www.e-taxonomy.eu/, visited on 09/29/2008.

[59] http://www.inspire-geoportal.eu/index.htm, visited on 09/29/2008.

[60] http://www.keytonature.eu/, visited on 09/29/2008.

Noël Conruyt, Didier Sébastien, Rémy Courdier, Daniel David, Nicolas Sébastien and Tiana Ralambondrainy
EA2525-IREMIA ICIHM & SMART Research Teams
2 rue Joseph Wetzel
97490 Sainte-Clotilde
Réunion - France
e-mail: {noel.conruyt, didier.sebastien, remy.courdier, daniel.david, nicolas.sebastien, tiana.ralambondrainy}@univ-reunion.fr

Whitestein Series in Software Agent Technologies and Autonomic Computing, 91–117
© 2009 Birkhäuser Verlag Basel/Switzerland

OSM: A Multi-Agent System for Modeling and Monitoring the Evolution of Oil Slicks in Open Oceans

Juan Manuel Corchado, Aitor Mata and Sara Rodríguez

Abstract. A multi-agent based prediction-system is presented in which the aim is to forecast the presence of oil slicks in a certain area of the open sea after an oil spill. In this case, the multi-agent architecture incorporates a prediction-system based on the CBR methodology, implemented in a series of interactive services, for modeling and monitoring the ocean water masses. The system's nucleus is formed by a series of deliberative agents acting as controllers and administrators for all the implemented services. The implemented services are accessible in a distributed way, and can be accessed even from mobile devices. The proposed system uses information such as sea salinity, sea temperature, wind, currents, pressure, number and area of the slicks, etc. obtained from various satellites. The system has been trained using data obtained after the Prestige accident. The Oil Spill Multi-Agent System (OSM) has been able to accurately predict the presence of oil slicks in the north-west of the Galician coast using historical data.

Keywords. Multi-Agent Systems, Oil spill, CBR agents, PCA.

1. Introduction

The response to minimize the environmental impact when an oil spill is produced must be precise, fast and coordinated. The use of contingency response systems can facilitate the planning and tasks assignation when organizing resources, especially when multiple people are involved.

When an oil spill is produced, the response to minimize the impact must be precise, fast and coordinated. In that kind of situations, where multiple people are involved, a flexible and distributed architecture is needed in order to develop effective contingency response systems.

One of the most important characteristics is the use of intelligent agents as the main components in employing a service oriented approach, focusing on distributing the majority of the systems' functionalities into remote and local services and applications. The architecture proposes a new and easier method of building distributed multi-agent systems, where the functionalities of the systems are not integrated into the structure of the agents, rather they are modeled as distributed services and applications that are invoked by the agents acting as controllers and coordinators.

Agents have a set of characteristics, such as autonomy, reasoning, reactivity, social abilities, pro-activity, mobility, organization, etc. which allow them to cover several needs for artificial intelligence environments [30], especially ubiquitous communication and computing and adaptable interfaces [8]. Agent and multi-agent systems have been successfully applied to several scenarios, such as education, culture, entertainment, medicine, robotics, etc. [6][25]. The characteristics of the agents make them appropriate for developing dynamic and distributed systems, as they possess the capability of adapting themselves to the users and environmental characteristics [14]. The continuous advancement in mobile computing makes it possible to obtain information about the context and also to react physically to it in more innovative ways . The agents in this architecture are based on the deliberative Belief, Desire, Intention (BDI) model [3], where the agents' internal structure and capabilities are based on mental aptitudes, using beliefs, desires and intentions. Nevertheless, modern developments need higher adaptation, learning and autonomy levels than pure BDI model [3]. This is achieved in new multi-agent architectures by modeling the agents' characteristics to provide them with mechanisms that allow solving complex problems and autonomous learning. Some of these mechanisms are Case-Based Reasoning (CBR) [1] and Case-Based Planning (CBP), where problems are solved by using solutions to similar past problems [6]. Solutions are stored in a case memory, which the mechanisms can consult in order to find better solutions for new problems. CBR and CBP mechanisms have been modeled as external services. Deliberative agents use these services to learn from past experiences and to adapt their behavior according the context.

Predicting the behavior of oceanic elements is a quite difficult task. In this case the prediction is related to external elements (oil slicks), and this makes the prediction even more difficult. An open ocean is a highly complex system that may be modeled by measuring different variables and structuring them together. Some of those variables are essential to predict the behavior of oil slicks. In order to predict the future presence of oil slicks in an area, it is obviously necessary to know their previous positions. That knowledge is provided by the analysis of satellite images, obtaining the precise position of the slicks.

The solution proposed in this article generates, for different geographical areas, a probability (between 0 and 1) of finding oil slicks after an oil spill. The OSM has been constructed using historical data and checked using the data acquired during the Prestige oil spill, from November 2002 to April 2003. Most of the data used to develop OSM has been acquired from the ECCO (*Estimating the Circulation and Climate of the Ocean*) consortium [17]. Position and size of the slicks has been obtained by treating SAR (*Synthetic Aperture Radar*) satellite images [18].

The proposed system uses a CBR structure to learn from past situations, and to generate solutions to new problems based on past solutions given to past problems. Past solutions are stored in the system in the *case base*. The cases contain information about the oil slicks (size and number) as well as atmospheric data (wind, current, salinity, temperature, ocean height and pressure). The OSM combines the efficiency of the CBR systems with artificial intelligence techniques in order to improve the results and to better generalize from past data. The results obtained approximate to the real process occurred in near 90% of the value of the main variables analyzed, which is a quite important approximation.

The OSM allows different users to work together but without sharing the same space. The multi-agent architecture divides the system in small pieces that work separately but coordinated. The different people involved in a contingency system like the described in this article can develop their specialized work being coordinated in the distance.

After an oil spill, it is necessary to determine whether an area is going to be contaminated or not. To conclude about the presence or not of contamination in an area, it is necessary to know the behavior of the slicks that are generated by the spill.

At first, position, shape and size of the oil slicks must be identified. One of the most precise ways to acquire that information is by using satellite images. SAR (*Synthetic Aperture Radar*) images are the most commonly used to automatically detect this kind of slicks [27]. Satellite images show certain areas where it seems to be nothing (e.g. zones with no waves) as oil slicks. Figure 1 shows a SAR image that displays a portion of the Galician west coast with black areas corresponding to oil slicks.

Figure 2 shows the interpretation of the SAR image after treating the data. SAR images make it possible to distinguish between normal sea variability and oil slicks. It is also important to make a distinction between oil slicks and look-alikes. Oil slicks are quite similar to quiet sea areas, so it is not always easy to discriminate between them. If there is not enough wind, the difference between the calm sea and the surface of an oil slick is less evident. This can lead to mistakes when trying to differentiate between a normal situation and an oil slick. This is a crucial aspect in this problem that can be automatically managed by computational tools [24]. Once the slicks are correctly identified, it is also crucial to know the atmospheric and maritime situation that is affecting the zone at the moment that is being analyzed. Information collected from satellites is used to obtain the atmospheric data needed. That is how different variables such as temperature, sea height and salinity are measured in order to obtain a global model that can explain how slicks evolve.

There are different ways to analyze, evaluate and predict situations after an oil spill. One approach is simulation [4], where a model of a certain area is

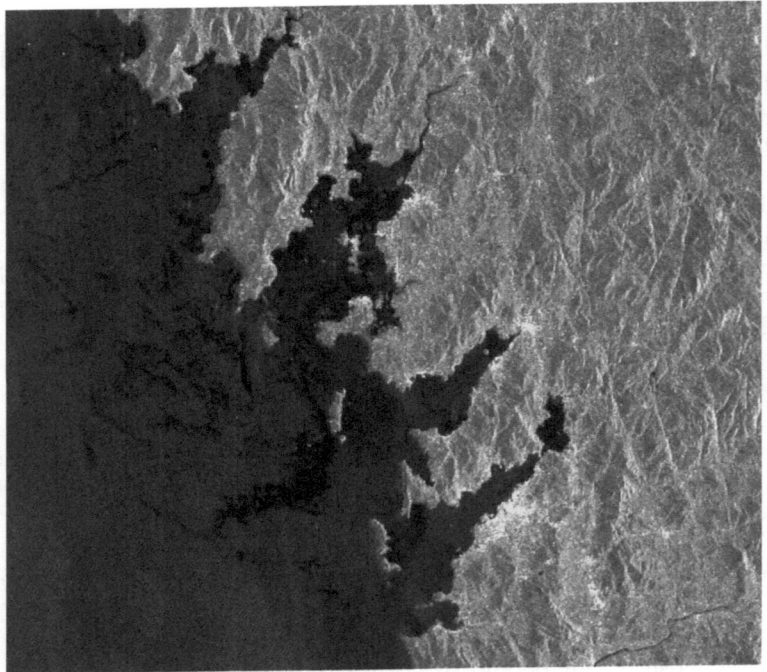

FIGURE 1. SAR image with an oil spill near the north-west coast of Spain.

created introducing specific parameters (weather, currents and wind) and working along with a forecasting system. Using simulations it is easy to obtain a good solution for a certain area, but it is quite difficult to generalize in order to solve the same problem in related areas or new zones. It is possible to replace the oil spill by drifters to obtain a trajectory model comparing the trajectory followed by the drifters with the already known oil slicks trajectories. If the drifters follow a similar trajectory as the one that followed the slicks, then a model can be created and there will be a possibility of creating more models in different areas. Another way of predicting oil slicks trajectories is studying previous cases for obtaining a trajectory model for a certain area [28]. One step over these solutions is the use of systems that – by combining a major set of elements – generate response models to solve the oil spill problem. A different point of view is given by complex systems that analyze large databases (environmental, ecological, geographical and engineering) using expert systems. This way, an implicit relation between problem and solution is obtained, but with no direct connection between past examples and current decisions. Nevertheless arriving at these kinds of solutions requires a great data mining effort. Once the oil spill is produced there should be contingency models for making a fast solution possible. Expert systems have also been used

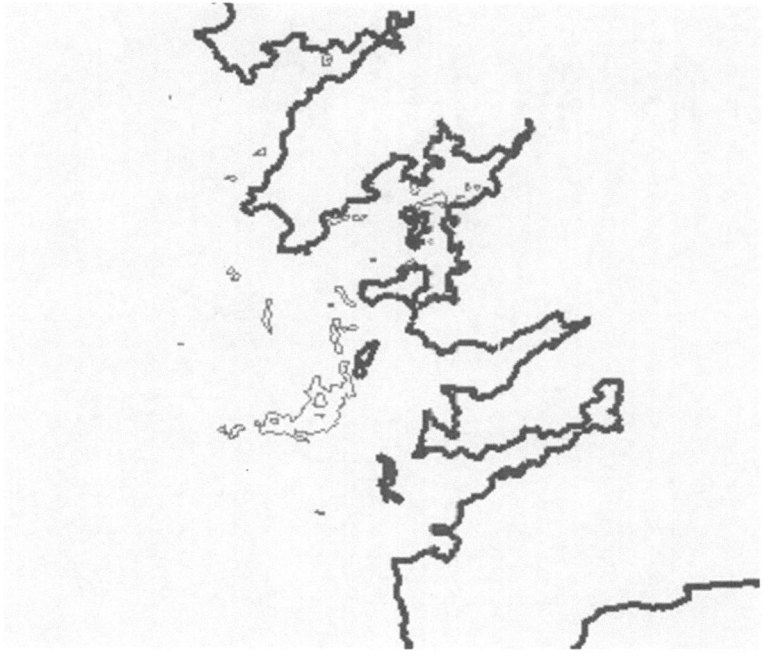

FIGURE 2. Interpretation of a SAR image done by the system.

for solving this problem. These systems use stored information from past cases as a repository where future applications will find structured information. The final objective of all these approaches is to provide decision support systems in order to enhance the response against oil spill situations. Different techniques have been used to achieve this objective, from fuzzy logic to negotiation with multi-agent systems. One of these techniques is Case-Based Reasoning which is described in the next section.

In this article, the oil spill problem is first presented, showing its difficulties and the possibilities of finding solutions to the problem. Then, the multi-agent architecture is described. Afterwards, OSM is explained, and last, the results are shown and also the future developments that can be achieved with the system.

2. A Multi-Agent Communication Architecture for Integrating Distributed Services

A multi-agent architecture has been developed to integrate the prediction-services. Because the architecture acts as an interpreter, the users can run applications and services programmed in virtually any language, but have to follow a communication protocol that all applications and services must incorporate. Another important

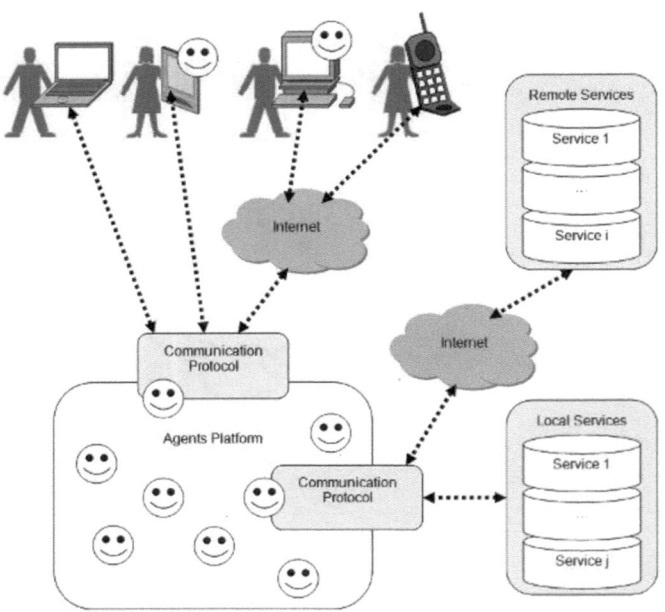

FIGURE 3. Framework basic schema.

functionality is that, thanks to the agents' capabilities, the systems developed can make use of reasoning mechanisms or learning techniques to handle services and applications according to context characteristics, which can change dynamically over time. Agents, applications and services can communicate in a distributed way, even from mobile devices. This makes it possible to use resources independent of their location. It also allows the starting or stopping of agents, applications, services or devices separately, without affecting the remaining resources, so the system has an elevated adaptability and capacity for error recovery.

Users can access the system through distributed applications, which run on different types of devices and interfaces (e.g. computers, cell phones, PDA). Figure 3 shows the basic scheme of the framework where all requests and responses are handled by the agents in the platform. The agents analyze all requests and invoke the specified services either locally or remotely. Services process the requests and execute the specified tasks. Then, services send back a response with the result of the specific task.

The presented framework is a modular multi-agent architecture, where services and applications are managed and controlled by deliberative BDI (Belief, Desire, Intention) agents [3, 15, 20, 29]. Deliberative BDI agents are able to cooperate, propose solutions on very dynamic environments, and face real problems, even when they have a limited description of the problem and few resources available. These agents depend on beliefs, desires, intentions and plan representations to solve problems [2, 11, 22]. Deliberative BDI agents build the core of the multi-agent communication system. There are different kinds of agents in the architecture, each one with specific roles, capabilities and characteristics. This fact facilitates the flexibility of the architecture in incorporating new agents. However, there are pre-defined agents that provide the basic functionalities of the architecture:

- **CommApp Agent**. This agent is responsible for all communications between applications and the platform. It manages the incoming requests from the applications to be processed by services. It also manages responses from services (via the platform) to applications. The CommApp Agent is always on listening mode. Applications send XML messages to the agent requesting a service, then the agent creates a new thread to start communication by using sockets. The agent sends all requests to the Manager Agent that processes the request. The socket remains open until a response to the specific request is sent back to the application using another XML message. All messages are sent to the Security Agent for their structure and syntax to be analyzed.
- **CommServ Agent**. It is responsible for all communications between services and the platform. The functionalities are similar to the CommApp Agent but backwards. This agent is always on listening mode waiting for responses of services. The Manager Agent signals to the CommServ Agent which service must be invoked. Then, the CommServ Agent creates a new thread with its respective socket and sends an XML message to the service. The socket remains open until the service sends back a response. All messages are sent to the Security Agent for their structure and syntax to be analyzed. This agent also periodically checks the status of all services to know whether they are idle, busy, or crashed.
- **Directory Agent**. It manages the list of services that can be used by the system. For security reasons [26], the list of services is static and can only be modified manually; however, services can be added, erased or modified dynamically. The list contains the information of all trusted available services. The name and description of the service, parameters required, and the IP address of the computer where the service is running are some of the information stored in the list of services. However, there is dynamic information that is constantly being modified: the service performance (average time to respond to requests), the number of executions, and the quality of the service. The latter data is very important, as it assigns a value between 0 and 1 to all services. All new services have a quality of service (QoS) value set to 1. This value decreases when the service fails (e.g. service crashes, no service found, etc.) or has a subpar performance compared to similar past executions. QoS is increased each time the service efficiently processes the tasks assigned. Information management is especially important because the data processed is very sensitive and personal. Thus, security must be a major concern. For this reason the multi-agent architecture does not implement a service discovery mechanism, requiring systems to employ only the specified

services from a trusted list of services. However, agents can select the most appropriate service (or group of services) to accomplish a specific a task.

- **Supervisor Agent**. This agent supervises the correct functioning of the other agents in the system. The Supervisor Agent periodically verifies the status of all agents registered in the architecture by sending ping messages. If there is no response, the Supervisor agent kills the agent and creates another instance of that agent.
- **Security Agent**. This agent analyzes the structure and syntax of all incoming and outgoing XML messages. If a message is not correct, the Security Agent informs the corresponding agent (CommApp or CommServ) that the message cannot be delivered. This agent also directs the problem to the Directory Agent, which modifies the QoS of the service where the message was sent.
- **Manager Agent**. Decides which agent must be called by taking into account the QoS and user preferences. Users can explicitly invoke a service, or can let the Manager Agent decide which service is best to accomplish the requested task. If there are several services that can resolve the task requested by an application, the agent selects the optimal choice. An optimal choice has a higher QoS and a better performance. The Manager Agent has a routing list to manage messages from all applications and services. This agent also checks if services are working properly. It requests the CommServ Agent to send ping messages to each service on a regular basis. If a service does not respond, the CommServ Agent informs the Manager Agent, which tries to find an alternate service, and informs the Directory Agent to modify the respective QoS.
- **Interface Agent**. This kind of agent was designed to be embedded in users' applications. Interface agents communicate directly with the agents in the architecture, so there is no need to employ the communication protocol, rather the FIPA ACL specification. The requests are sent directly to the Security Agent, that analyzes the requests and sends them to the Manager Agent. The rest of the process follows the same guidelines for calling any service. These agents must be simple enough to allow them to be executed on mobile devices, such as cell phones or PDAs. All high demand processes must be delegated to services.

In the next section, the contingency response system to face oil slick situations is presented, explaining how the multi-agent architecture is integrated with a CBR system in order to obtain a flexible and distributed structure.

3. OSM: A Hybrid Multi-Agent System for Contingency Response in Oil Spill Situations

CBR has already been used to solve maritime problems [7] in which different oceanic variables were involved. In this case, the data collected from different

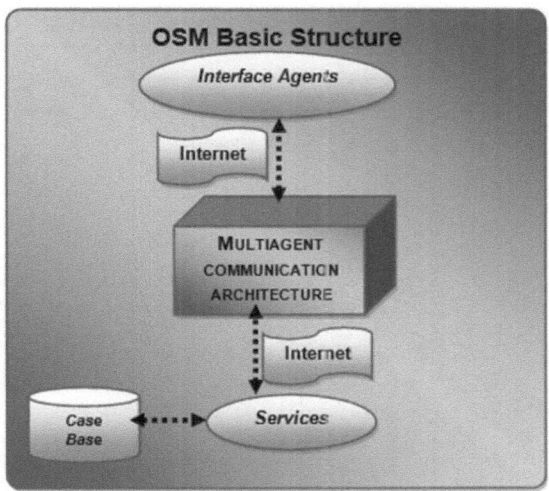

FIGURE 4. OSM basic structure.

observations from satellites is processed and structured as cases. The cases are the key to obtain solutions to future problems through a CBR system.

Figure 4 shows the basic structure of the OSM system, where the interfaces agents are connected to the services through the multi-agent architecture. The interface agents represent the different roles the users can perform to interact with the system. The services are the different phases of the CBR cycle that are requested by the users. One user may only need to introduce information in the system, while expert users can be requested from the system to confirm the predictions generated. OSM is a hybrid multi-agent system, combining the main capabilities of multi-agent systems with the prediction and generalization power of the CBR systems.

The functionalities of the system can be accessed using different interfaces for PCs and PDAs (Personal Digital Assistants) where users can interact with the system by introducing data, requesting a prediction or revising a solution generated by the system. Figure 5 shows the main graphical user interface of OSM. The interface shows a set of parameters, the oceanic area visualization with oil slicks and a squared area to be analyzed.

Oil slicks are mainly detected using SAR images. Those images are processed and transformed to be used by the system. Oceanic, meteorological and oil spill related data is stored in the system in order to generate future predictions. The data used to train the system has been obtained after the Prestige accident, between November 2002 and April 2003, in a specific geographical area at the west of the Galician coast (longitude between 14 and 6 degrees west and latitude between 42 and 46 degrees north). Table 1 shows the basic structure of a case. The variables

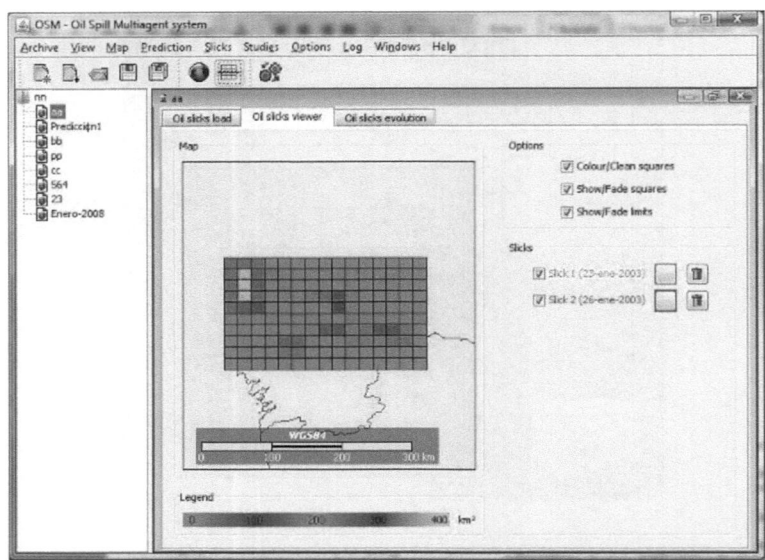

FIGURE 5. Graphical user interface of OSM.

can be geographical (longitude and latitude), temporal (date of the case), atmospheric (wind, current, sea height, bottom pressure, salinity and temperature) and directly related with the problem (number and area of the slicks).

All information is stored in the case base and the OSM is ready to predict future situations. A problem situation must be introduced in the system for generating a prediction. Then, the most similar cases to the current situation are retrieved from the case base. Once a collection of cases are chosen from the case base, they are used for generating a new solution to the current problem. Growing Radial Basis Functions Networks [16] are used in OSM for combining the chosen cases in order to obtain the new solution.

The OSM determines the probability of finding oil slicks in a certain area. To do so, the OSM divides the area to be analyzed in squares of approximately half a degree side for generating a new prediction. Then, the system determines the amount of slicks in each square. The squares are colored with different gradation depending on the quantity of oil slicks calculated.

Within the case base there is a temporal relationship between a case and its future situation. A square, with all the values of the different variables can be related with the same square but in the next temporal situation. This relationship will provide the internal mechanism used to generalize and to train the GRBF network that will generate the prediction.

Figure 6 shows the interpretation of a series of slicks. The squared areas are those that will be analyzed by the system. First, the slicks corresponding to

Variable	Definition	Unit
Longitude	Geographical longitude	Degree
Latitude	Geographical latitude	Degree
Date	Day, month and year of the analysis	dd/mm/yyyy
Sea Height	Height of the waves in open sea	m
Bottom pressure	Atmospheric pressure in the open sea	Newton/m^2
Salinity	Sea salinity	ppt (parts per thousand)
Temperature	Celsius temperature in the area	C
Area of the slicks	Surface covered by the slicks present in the area	Km^2
Meridional Wind	Meridional direction of the wind	m/s
Zonal Wind	Zonal direction of the wind	m/s
Wind Strength	Wind strength	m/s
Meridional Current	Meridional direction of the ocean current	m/s
Zonal Current	Zonal direction of the ocean current	m/s
Current Strength	Ocean current strength	m/s

TABLE 1. Variables that define a case.

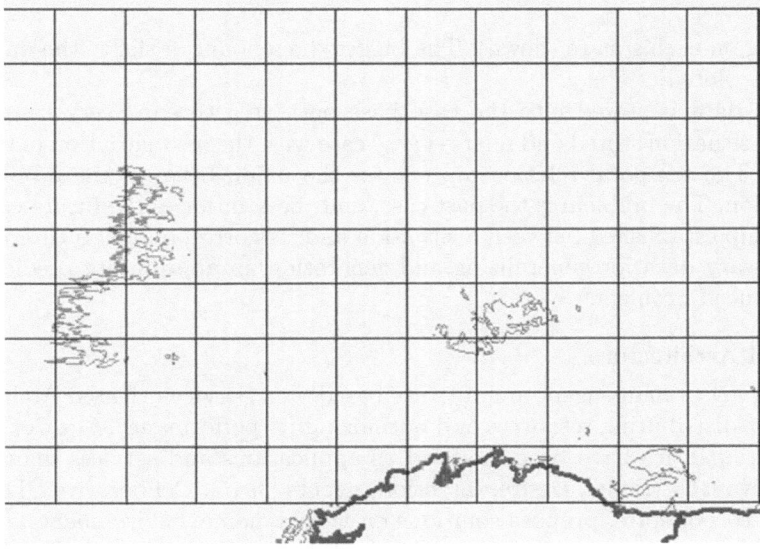

FIGURE 6. Division of the analyzed area into squares with slicks inside.

different days are colored in different colors. Then, in Figure 7 can be seen how the squared zones are colored in different intensity depending on the amount of slicks

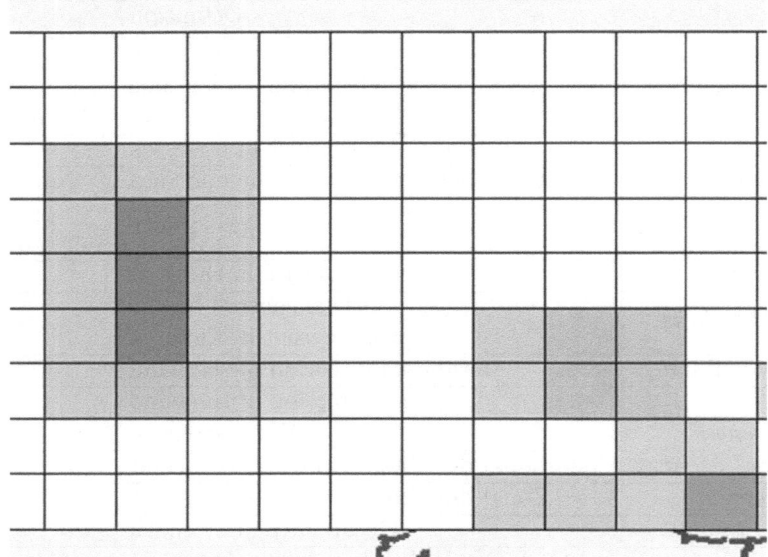

FIGURE 7. Interpretation of the amount of slicks in an area.

appearing on each square (down). The bigger the amount of slicks, the darker the square is colored.

The data is stored into the case base once structured. Every case has its temporal situation stored and relates every case with the next situation in the same position. The temporal relationship creates the union between the problem and the solution. The problem is the past case, and the solution is the future case. The relationship established between a situation and its corresponding future provides the necessary data for generalizing and generating an appropriate prediction for an introduced problem.

3.1. OSM Architecture

OSM employs a multi-agent architecture based on a Service-oriented Architecture (SOA) for distributing resources and optimizing its performance. Most of the system functionalities have been modeled as applications and services managed by deliberative BDI (Belief, Desire, Intention) agents [3, 15]. Deliberative BDI agents are able to cooperate, propose solutions on very dynamic environments, and face real problems, even when they have a limited description of the problem and few resources available. These agents depend on beliefs, desires, intentions and plan representations to solve problems [11].

There are four basic blocks in OSM: Applications, Services, Agents Platform and Communication Protocol. These blocks provide the complete system functionalities:

- **Applications**. These represent all the programs that users can use to exploit the system functionalities. Applications are dynamic, reacting differently according to the particular situations and the services invoked. They can be executed locally or remotely, even on mobile devices with limited processing capabilities, because computing tasks are largely delegated to the agents and services.
- **Services**. These represent the activities that the architecture offers. They are the bulk of the functionalities of the system at the processing, delivery and information acquisition levels. Services are designed to be invoked locally or remotely. Services can be organized as local services, web services, GRID services, or even as individual stand-alone services. Services can make use of other services to provide the functionalities that users require. The OSM has a flexible and scalable directory of services, so they can be invoked, modified, added, or eliminated dynamically and on demand. It is absolutely necessary that all services follow a communication protocol to interact with the rest of the components.
- **Agents Platform**. This is the core of the system, integrating a set of agents, each one with special characteristics and behavior. An important feature in this architecture is that the agents act as controllers and administrators for all applications and services, managing the adequate functioning of the system, from services, applications, communication and performance to reasoning and decision-making. In OSM, services are managed and coordinated by deliberative BDI agents. The agents modify their behavior according to the users' preferences, the knowledge acquired from previous interactions, as well as the choices available to respond to a given situation.
- **Communication Protocol**. This allows applications and services to communicate directly with the Agents Platform. The protocol is completely open and independent of any programming language. This protocol is based on SOAP specification to capture all messages between the platform and the services and applications [5]. Services and applications communicate with the *Agents Platform* via SOAP messages. A response is sent back to the specific service or application that made the request. All external communications follow the same protocol, while the communication among agents in the platform follows the FIPA Agent Communication Language (ACL) specification. This is especially useful when applications run on limited processing capable devices (e.g. cell phones or PDAs). Applications can make use of agents platforms to communicate directly (using FIPA ACL specification) with the agents in the OSM, so while the communication protocol is not needed in all instances, it is absolutely required for all services.

Agents, applications and services in the OSM can communicate in a distributed way, even from mobile devices. This makes it possible to use resources independent of their location. It also allows the starting or stopping of agents, applications, services or devices separately, without affecting the remaining resources,

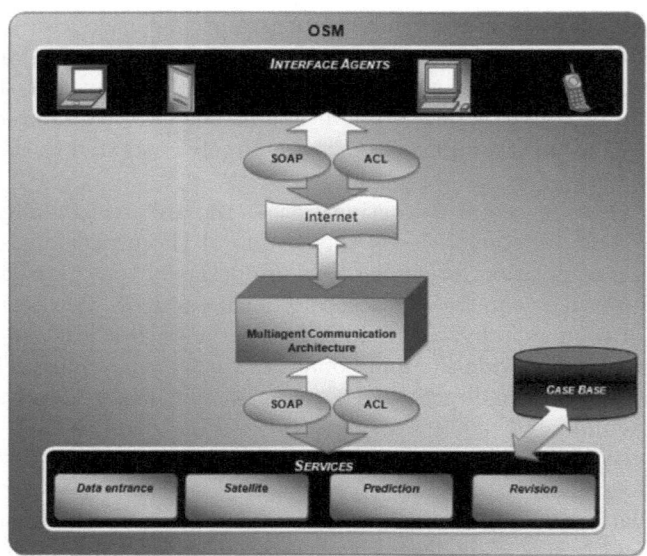

FIGURE 8. OSM extended structure.

so the system has an elevated adaptability and capacity for error recovery. Users can access to OSM functionalities through distributed applications that run on different types of devices and interfaces (e.g. computers, PDA). Figure 8 shows the structure of OSM. As can be seen, most of the functionalities, including the CBR system, have been modeled as services and applications. Thus, each service can be performed on demand and can also be replicated to respond multiple requests.

Interface Agents are a special kind of agents in OSM designed to be embedded in users' applications. These agents are simple enough to allow them to be executed on mobile devices, such as cell phones or PDAs because all high demand processes are delegated to services. OSM defines three different *Interface Agents*:

1. *Input Agent.* It is the agent that sends the information introduced by the users to OSM. Once the data have reached the system, it is structured into the case base. This interface agent is used by users that have visualized an oil slick, in order to introduce the data related with that slick into the system. The Input Agent interface is used by the users to introduce the data. The main parameters to identify the slick are its position, in terms of longitude and latitude, the surface covered by the slick, and the distance of that slick to the coast. Once the basic information about the slick has been sent to the system, the OSM recovers satellite information about the ocean and the meteorological conditions in the area to create a case from the slick and geographical information.

2. *Prediction Agent.* When a user wants to request a prediction from the OSM, this is the agent used to do so. In the interface of the agent, the user can define the area to be analyzed, the size of the squares to be transformed into cases and, if there are previous informations stored in the system, the existing slicks to be considered to generate the prediction.

3. *Revision Agent.* When a prediction is generated by OSM, the system can automatically verify the correction of the proposed solution. But, if there are revision experts available, it also requests an expert for a revision. The users receive the proposed solution and enough data to validate the solution for the current problem.

The OSM also defines three different services that perform all tasks that the users may demand from the system. All requests and responses are handled by the agents. The requests are analyzed and the specified services are invoked either locally or remotely. Services process the requests and execute the specified tasks. Then, services send back a response with the result of the specific task. In this way, the agents act as interpreters between applications and services in OSM.

Next, the main services defined in OSM are explained, following the main phases of the CBR cycle.

3.2. Prediction System

The OSM is a Contingency Response system for oil spills conceived as a multi-agent system which core working structure follows the Case-Based Reasoning methodology. The different services implemented by the OSM system cover the four main phases of the OSM cycle, and also the pre-processing stage, covered by the *Data Input Service* covers. The retrieval and reuse phases are implemented in the *Prediction Generation Service* that generates a prediction after a problem description is introduced in the system by an user. The *Revision Service* covers the *revision* phase, and may require the confirmation of an expert, to validate the correction of the solution proposed. The final stage of the CBR cycle, the *retention* phase, is also implemented in the *Data Input Service*, as described in the following.

Data Input Service. When data about an oil slick is introduced in the system, the OSM must complete the information about the area including atmospheric and oceanic information: temperature, salinity, bottom pressure, sea height. All that complementary data is collected from satellite services that offer on-line and in real time that precise information. With all this information, the case is created and introduced in the case base.

Historical data collected from November 2002 to April 2003 has been used to create the case base of OSM. As explained before, cases are formed by a series of variables. Principal Components Analysis (PCA) [9] can reduce the number of those variables, and then the system stores the value of the principal components, which are related to the original variables that define a case. PCA has been previously used to analyze oceanographic data and it has proved to be a consistent technique when trying to reduce the number of variables [21]. OSM uses Fast Iterative Kernel PCA (FIKPCA) which is an evolution of PCA [12]. This technique reduces the number of variables in a set by eliminating those that are linearly dependent, and it is quite faster than the traditional PCA.

To improve the convergence of the Kernel-Hebbian-Algorithm used by Kernel PCA, FIK-PCA sets η_t proportional to the reciprocal of the estimated eigenvalues. Let $\lambda_t \in \Re^r_+$ denote the vector of eigenvalues associated with the current estimate of the first r eigenvectors to which the algorithm converges. The new KHA algorithm sets the $i^t h$ component of η_t to the files:

$$[\eta_t]_i = \frac{1}{[\lambda_t]_i} \frac{\tau}{t + \tau} \eta_0 \qquad (3.1)$$

The final variables are, obviously, linearly independent and are formed by a combination of the previous variables. The values of the original variables can be recovered by doing the inverse calculation to the one produced to obtain the new variables. The variables that are less used in the final stored variables are those whose values suffer less changes during the periods of time analyzed (salinity, temperature and pressure do not change from one day to another, then, they can be ignored considering that the final result does not depend on them). Once the FIKPCA is applied, the number of variables is reduced to three, having the following distribution:

```
Variable_1: -0,560 * long - 0,923*lat + 0,991*s_height +
0,919*b_pressure + 0,992*salinity + 0,990*temp -
0,125*area_of_slicks + 0,80*mer_wind + 0,79*zonal_wind +
0,123*w_strenght + 0,980*mer_current + 0,980*zonal_current
+ 0,980*c_strength

Variable_2: 0,292*long - 0,081*lat - 0,010*s_height -
0,099*b_pressure - 0,011*salinity - 0,013*temp -
0,021*area_of_slicks + 0,993*merl_wind + 0,993*zonal_wind
+ 0,989*w_strenght - 0,024*mer_current - 0,024*zonal_current
- 0,024*c_strength

Variable_3: 0*long - 0,072*lat + 0,009*s_height +
0,009*b_pressure + 0,009*salinity + 0,009*temp +
0,992*area_of_slicks + 0,006*mer_wind + 0,005*zonal_wind
+ 0,005*w_strenght - 0,007*mer_current - 0,007*zonal_current
- 0,007*c_strength
```

After applying FIKPCA, the historical data is stored in the case base, and is used to solve future problems using the rest of the CBR cycle. Storing the principal components instead of the original variables implies reducing the amount of memory necessary to store the information in about a sixty per cent which is more important as the case base grows. The reduction of the number of variables considered also implies a faster recovery from the case base.

When introducing the data into the case base, Growing Cell Structures (GCSs) [10] are used. GCSs can create a model from a situation organizing the different cases by their similarity. If a two-dimensional representation is chosen to explain this technique, the most similar cells (i.e. cases) are near to each other. If there is a relationship between the cells, they are grouped together, and this grouping characteristic helps the CBR system to recover the similar cases in the next phase. When a new cell is introduced in the structure, the closest cells move towards the new one, changing the overall structure of the system. The weights of the winning cell w_c and its neighbors w_n are changed. The terms ϵ_c and ϵ_n represent the learning rates for the winner and its neighbors, respectively. x represents the value of the input vector.

$$w_c\,(t+1) = w_c\,(t) + \epsilon_c\,(x - w_t) \tag{3.2}$$

$$w_n\,(t+1) = w_n\,(t) + \epsilon_n\,(x - w_n) \tag{3.3}$$

The pseudocode of the GCS insertion process is shown below:

1. The most similar cell to the new one is found.
2. The new cell is introduced in the middle of the connection between the most similar cell and the least similar to the new one.
3. Direct neighbors of the closest cell change their values by approximating to the new cell and specified percentage of the distance between them and the new cell.

Once the case base has stored the historical data and the GCS has learned from the original distribution of the variables, the system is ready to receive a new problem.

When a new problem comes to the system, GCSs are used once again. The stored GCS behaves as if the new problem would be stored in the structure and finds the most similar cells (cases in the CBR system) to the problem introduced in the system. In this case, the GCS does not change its structure, because it has being used to obtain the most similar cases to the introduced problem. Only in the retain phase the GCS changes again, introducing the proposed solution if it is correct.

Prediction Generation Service. When a prediction is requested by a user, the system starts recovering from the case base the most similar cases to the problem proposed. Then, it creates a prediction using artificial neural networks.

The similarity between the new problem and the cases is determined by the GCS. Every element in the GCS has a series of values (every value corresponds to one of the principal components created after the PCA analysis). The distance between elements is a multi-dimensional distance where all the elements are considered to establish the distance between cells. After obtaining the most similar cases from the case base, the cases are used in the next phase. The most similar cases stored in the case base will be used to obtain an accurate prediction according to the previous solutions related with the selected cases.

Once the most similar cases are recovered from the case base, they are used to generate the solution. The prediction of the future probability of finding oil slicks in an area is generated using an artificial neural network with a hybrid learning system. An adaptation of Radial Basis Functions Networks (RBFNs) is used to obtain that prediction [13]. The chosen cases are used to train the artificial neural network. Radial Basis Function Networks have been chosen because of the reduction of the training time compared to other artificial neural network systems, such as Multilayer Perceptrons. In this case, the network is trained in every analysis using only the cases selected from the case base.

Growing RBFNs [23] are used to obtain the predicted future values corresponding to the proposed problem. This adaptation of the RBFNs allows the system to grow during training gradually increasing the number of elements (prototypes) that play the role of the centers of the radial basis functions. The creation of the Growing RBFN must be made automatically which implies an adaptation of the original GRBF system. The error for every pattern is defined by:

$$e_i = l/p * \sum_{k=1}^{p} ||t_{ik} - y_{ik}||, \qquad (3.4)$$

where t_{ik} is the desired value of the k_{th} output unit of the i_{th} training pattern, y_{ik} the actual values of the kth output unit of the i_{th} training pattern.

The pseudo code of the Growing RBF process id described next:

1. Calculate the error, e_i (3.4) for every new possible prototype.
        a. If the new candidate does not belong to the chosen ones and the error calculated is less than a threshold error, then the new candidate is added to the set of accepted prototypes.
        b. If the new candidate belongs to the accepted ones and the error is less than the threshold error, then modify the weights of the neurons in order to adapt them to the new situation.
2. Select the best prototypes from the candidates.
        a. If there are valid candidates, create a new cell centered on it.
        b. Else, increase the iteration factor. If the iteration factor comes to the 10% of the training population, freeze the process.

3. Calculate global error and update the weights.
        a. If the results are satisfactory, end the process. If
    not, go back to step 1.

Once the GRBF network is created, it is used to generate the solution to the proposed problem. The solution will be the output of the network using as input data the selected cases from the case base.

Revision Service. After generating a prediction, the system needs to validate its correction. The system can also query an expert user to confirm the automatic revision previously done. The prediction is shown to the users in a similar way the slicks are interpreted by the OSM. A set of squared colored areas appear. The intensity of the color corresponds to the probability of finding oil slicks in that area. The areas colored with a higher intensity are those with the highest probability of finding oil slicks in them. In this visual approximation, the user can check if the solution is adequate. The system also provides an automatic method of revision that must be also checked by an expert user that confirms the automatic revision.

Explanations are a recent revision methodology used to check the correction of the solutions proposed by CBR systems [19]. Explanations are a kind of justification of the solution generated by the system. To obtain a justification to the given solution, the cases selected from the case base are used again. As explained before, we can establish a relationship between a case and its future situation. If we consider the two situations defined by a case and the future situation of that case as two vectors, we can define a distance between them, calculating the evolution of the situation in the considered conditions. That distance is calculated for all the cases retrieved from the case base as similar to the problem to be solved. If the distance between the proposed problem and the solution given is not greater than the average distances obtained from the selected cases, then the solution is a good one, according to the structure of the case base. Next, the explanation pseudo code is showed:

1. For every selected case in the retrieval phase, the distance
   between the case and its solution is calculated.
2. The distance between the proposed problem and the proposed
   solution is also calculated.
3. If the difference between the distance of the proposed solution
   and those of the selected cases is below a certain threshold
   value, then the solution is considered as a valid one.
4. If not, the user is informed and the process goes back to the
   retrieval phase, where new cases are selected from the case
   base.
5. If, after a series of iterations, the system does not produce a
   good enough solution, then the user is asked to consider the
   acceptance of the best of the generated solutions.

The distances are calculated considering the sign of the values, not using its absolute value. This decision is justified by the fact that it is not the same to move to the north than to the south, even if the distance between two points is the same. If the prediction is considered as correct, it is stored in the case base, and it can then be used in next predictions to obtain new solutions.

If the proposed prediction is accepted, it is considered as a good solution to the problem and can be stored in the case base in order to solve new problems. It will have the same category as the historical data previously stored in the system.

When inserting a new case in the case base, Fast Iterative Kernel PCA is used for reducing the number of variables used and adapting the data generated by the system. The adaptation is done by changing the original variables into the principal components previously chosen by the system. The internal structure of the case base also changes when a new case is introduced. The GCS system related to the case base structure controls its growth. The GCS system grows and improves its capability of generating good results as new knowledge is introduced in the system.

4. Results

The OSM uses different artificial intelligence techniques to cover and solve all the phases of the CBR cycle. Fast Iterative Kernel Principal Component Analysis is used to reduce the number of variables stored in the system, getting about a 60% reduction in the size of the case base. This adaptation of the PCA also implies a faster recovery of cases from the case base (more than 7% faster than storing the original variables).

To obtain a prediction using the cases recovered from the case base, Growing Radial Basis Function Networks have been used. This evolution of the RBF networks implies a better adaptation to the structure of the case base, which is organized using Growing Cell Structures. The results using Growing RBF networks instead of simple RBF networks are about 4% more accurate, which is a good improvement.

Evaluations show that the system can predict in the conditions already known, showing better results than previously used techniques. The use of a combination of techniques integrated in the CBR structure makes it possible to obtain better resulst than using the CBR alone (17% better), and also better than using the techniques isolated (neural networks), without the integration feature produced by the CBR (11% better). A resume of all these improvements can be seen in Figure 9.

The predicted situation was contrasted with the actual future situation. The future situation was known, as long as historical data was used to develop the system and also to test the correction of it. The proposed solution was, in most of the variables, close to 90% of accuracy. For every problem defined by an area and its variables, the system offers 9 solutions (i.e. the same area with its proposed

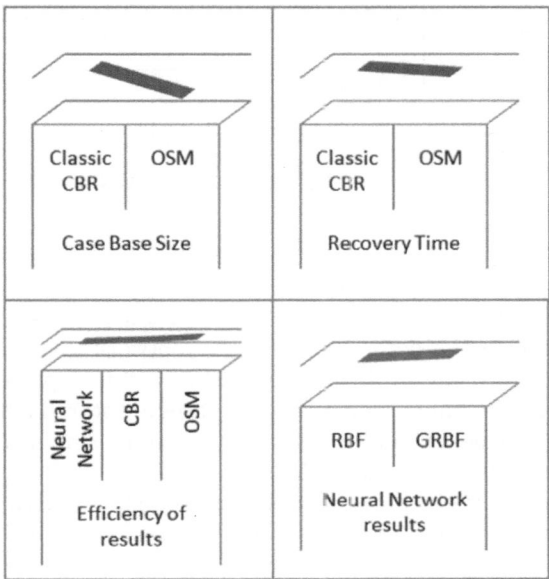

FIGURE 9. Resume of the improvement of the results obtained with OSM.

Number of cases	RBF	CBR	RBF + CBR	OSCBR
100	45%	39%	42%	43%
500	48%	43%	46%	46%
1000	51%	47%	58%	64%
2000	56%	55%	65%	72%
3000	59%	58%	68%	81%
4000	60%	63%	69%	84%
5000	63%	64%	72%	87%

TABLE 2. Percentage of good predictions obtained with different techniques.

variables and the eight closest neighbors). This way of prediction is used in order to clearly observe the direction of the slicks which can be useful in order to determine the coastal areas that will be affected by the slicks generated after an oil spill.

Table 2 shows a summary of the results obtained after comparing different techniques with the results obtained using OSM. The table shows the evolution of the results along with the increase of the number of cases stored in the case base. All the techniques analyzed improve the results while increasing the number of cases stored. Having more cases in the case base, makes it easier to find cases that are similar to the proposed problem and then, the solution can be more

.	RBF	CBR	RBF + CBR	OSCBR
RBF				
CBR	*			
RBF + CBR	=	=		
OSCBR	*	*	*	

TABLE 3. Multiple comparison procedure among different techniques.

accurate. The *RBF* column represents a simple Radial Basis Function Network that is trained with all the data available. The network gives an output that is considered a solution to the problem. The *CBR* column represents a pure CBR system, with no other techniques included; the cases are stored in the case base and recovered considering the Euclidean distance. The most similar cases are selected and after applying a weighted mean depending on the similarity of the selected cases with the inserted problem, a solution *s* proposed. The *RBF + CBR* column corresponds to the possibility of using a RBF system combined with CBR. The recovery from the CBR is done by the Manhattan distance and the RBF network works in the reuse phase, adapting the selected cases to obtain the new solution. The results of the *RBF+CBR* column are normally better than those of the *CBR*, mainly because of the elimination of useless data to generate the solution. Finally, the *OSM* column shows the results obtained by the proposed system, obtaining better results than the three previous analyzed solutions.

Table 3 shows a multiple comparison procedure (Mann-Whitney test) used to determine which models are significantly different from the others. The asterisk (*) indicates that these pairs show statistically significant differences at the 99.0% confidence level. OSM presents statistically significant differences with the rest of the models. The proposed solution does not generate a trajectory, but a series of probabilities in different areas, what is far more similar to the real behavior of the oil slicks.

Several tests have been done to compare the overall performance of OSM. The tests consisted of a set of requests delivered to the *Prediction Generation Service* (PGS) which in turn had to generate solutions for each problem. There were 50 different data sets, each one with 10 different parameters. The data sets were introduced into the PGS through a remote PC running multiple instances of the *Prediction Agent*. The data sets were divided in five test groups with 1, 5, 10, 20 and 50 data sets, respectively. There was one *Prediction Agent* for each test group. 30 runs for each test group were performed. Several data have been obtained from these tests, notably the average time to accomplish the solutions, the number of crashed agents, and the number of crashed services. First, all tests were performed with only one Prediction Service running in the same workstation on which the system was running. Then, five Prediction Services were replicated also in the same workstation. For every new test, the case base of the PGS was deleted in order

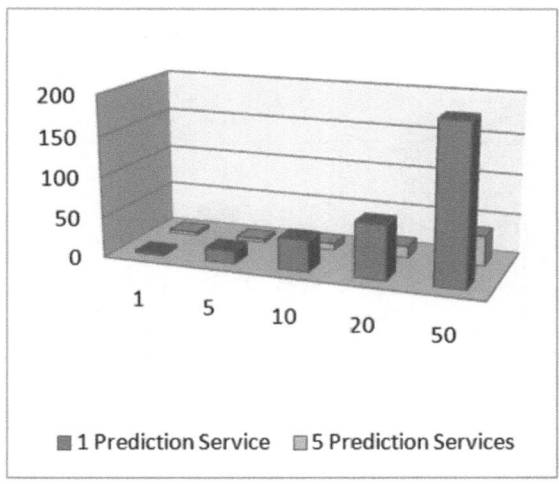

FIGURE 10. Average time needed to generate all solutions.

to avoid a learning capability, thus requiring the service to accomplish the entire prediction process.

Figure 10 shows the average time needed by the OSM for generating all solutions (predictions) for each test group. As can be seen, the time increases exponentially when there is only one PGS running. This is because the service must finish a request to start the next one. Hence, for the last test group (50 data sets) the service was overcharged. On the other hand, with five replicated services, the system can distribute the requests among these services and optimize the overall performance. The system performed slightly faster when processing a single request, but the performance was constantly reduced when more requests were sent to the service.

Figure 11 shows the average number of crashed agents and services during all tests. As can be seen, with only one PGS available, the OSM is far more unstable. This is because the PGS had to perform all requests by itself. It is important to notice that when the PGS crashed, more agents crashed because they were always waiting for the service response. For example, when processing 50 data sets, the last agent had to wait almost 200 seconds to receive the response. These data demonstrate that a distributed approach provides a higher ability to recover from errors.

5. Conclusions and Future Work

In this article, the OSM system has been explained. It is a new solution for predicting the presence of oil slicks in a certain area after an oil spill.

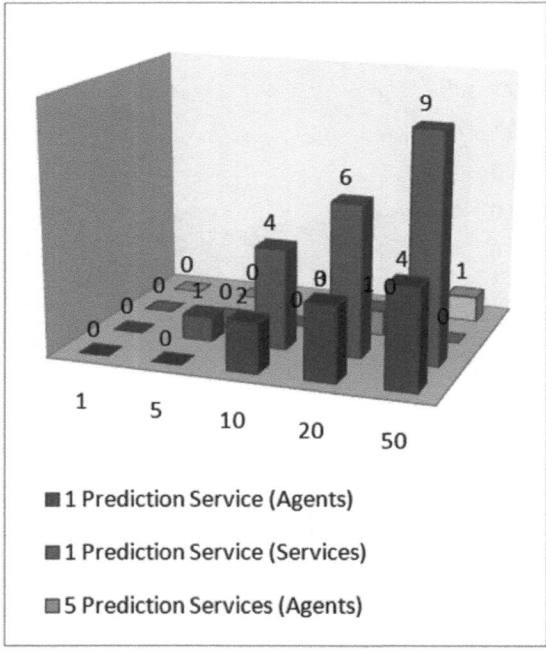

FIGURE 11. Average number of crashed agents.

The OSM represents a combination of a distributed multi-agent architecture with a prediction-system based on Case-Based Reasoning. The arrangement of those two methodologies allows the OSM to be able to interact with different users at the same time, generating solutions to new problems using solutions of past problems.

The distribution has been effective, permitting the users to interact with the system in different ways, depending on the needs of the users or on the requirements of the system.

CBR represents the predicting part of the system. It has proved to generate consistent results if enough data is available. The structure of the CBR methodology has been divided into services in order to adapt its way of working to the inner structure of the multi-agent architecture.

Generalization must be done in order to improve the system. Applying the methodology explained before to diverse geographical areas will make the results even better, being able to generate good solutions in more different situations.

Although these tests have provided us with very useful data, it is necessary to continue developing and enhancing OSM. The results also demonstrate that a distributed architecture is adequate for building complex systems and exploiting composite services, in this case OSM.

Acknowledgment

This work has been supported by the Spanish Ministry of Science and Technology project TIN2006-14630-C03-03 and the JCyL project SA071A08.

References

[1] Aamodt, A. and Plaza, E., *Case-Based Reasoning: foundational Issues*, Methodological Variations, and System Approaches, AI Communications, **7** (1) (1994), 39–59.

[2] Bratman, M.E., *Intentions, plans and practical reason*, Harvard University Press, Cambridge, MA, USA. (1987)

[3] Bratman, M.E., Israel, D. and Pollack, M.E., *Plans and resource-bounded practical reasoning*, Computational Intelligence, 4 (1988), 349–35.

[4] Brovchenko, I., Kuschan, A., Maderich, V. and Zheleznyak, M., *The modelling system for simulation of the oil spills in the Black Sea*, 3rd EuroGOOS Conference: Building the European capacity in operational oceanography., **192** (2002).

[5] Cerami, E., *Web Services Essentials Distributed Applications with XML-RPC, SOAP, UDDI & WSDL*, O'Reilly & Associates, Inc.

[6] Corchado, J.M., Bajo, J., De Paz, Y. and Tapia, D.I., *Intelligent Environment for Monitoring Alzheimer Patients*, Agent Technology for Health Care, Decision Support Systems, In Press.

[7] Corchado, J.M. and Fdez-Riverola, F., *FSfRT: Forecasting System for Red Tides*, Applied Intelligence, **21** (2004), 251–264.

[8] Chen, Y.M. and Wang, S.C. , *Framework of agent-based intelligence system with two-stage decision-making process for distributed dynamic scheduling*, Applied Soft Computing Journal, **7** (1) (2007), 229-245.

[9] Dunteman, G.H., *Principal Components Analysis*. Newbury Park, California (1989).

[10] Fritzke, B., *Growing cell structures-a self-organizing network for unsupervised and supervised learning*, Neural Networks, **7** (9) (1994), 1441–1460.

[11] Georgeff, M. and Rao, A. , *Rational software agents: from theory to practice*, In Jennings, N.R. y Wooldridge, M.J. (eds), Agent Technology: Foundations, Applications, and Markets. Springer-Verlag New York, Secaucus, NJ, (1998), 139–160.

[12] Gunter, S., Schraudolph, N.N. and Vishwanathan, S.V.N., *Fast Iterative Kernel Principal Component Analysis*, Journal of Machine Learning Research, **8** (2007), 1893–1918.

[13] Haykin, S.,*Neural networks*. Prentice Hall Upper Saddle River, NJ (1999).

[14] Jayaputera, G.T., Zaslavsky, A.B. and Loke, S.W.,*Enabling run-time composition and support for heterogeneous pervasive multi-agent systems*. Journal of Systems and Software,80(12) (2007), 2039–2062.

[15] Jennings, N.R. and Wooldridge, M.,*Applying agent technology*. Applied Artificial Intelligence,9(4) (1995), 351–361.

[16] Karayiannis, N.B. and Mi, G.W.,*Growing radial basis neural networks: merging supervised andunsupervised learning with network growth techniques*, Neural Networks, IEEE Transactions on, **8** (6) (1997), 1492–1506.

[17] Menemenlis, D., Hill, C., Adcroft, A., Campin, J.M., et al., *NASA Supercomputer Improves Prospects for Ocean Climate Research*, EOS Transactions, **86** (9) (2005), 89–95.

[18] Palenzuela, J.M.T., Vilas, L.G. and Cuadrado, M.S., *Use of ASAR images to study the evolution of the Prestige oil spill off the Galician coast*, International Journal of Remote Sensing, **27** (2006) (10), 1931–1950.

[19] Plaza, E., Armengol, E. and Ontañón, S., *The Explanatory Power of Symbolic Similarity in Case-Based Reasoning*, Artificial Intelligence Review, **24** (2) (2005), 145–161.

[20] Pokahr, A., Braubach, L. and Lamersdorf, W., *Jadex: Implementing a BDI-Infrastructure for JADE Agents*, In EXP - in search of innovation (Special Issue on JADE) (2003), 76–85.

[21] Preisendorfer, R.W., *Principal Component Analysis in Meteorology and Oceanography*, Development in atmospheric science (1988).

[22] Rao, A.S. and Georgeff, M.P., *BDI agents: from theory to practice*, In Proceedings of the First International Conference on Multi-Agents Systems (ICMAS'95). San Francisco, CA, USA. (1995).

[23] Ros, F., Pintore, M. and Chrétien, J.R., *Automatic design of growing radial basis function neural networks based on neighboorhood concepts*, Chemometrics and Intelligent Laboratory Systems, **87** (2) (2007), 231–240.

[24] Ross, B.J., Gualtieri, A.G., Fueten, F., Budkewitsch, P., et al., *Hyperspectral image analysis using genetic programming*, Applied Soft Computing, **5**(2) (2005), 147–156.

[25] Schn, B., O'Hare, G.M.P., Duffy, B.R., Martin, A.N., et al., *Agent Assistance for 3D World Navigation*, Lecture Notes in Computer Science, **3661**(2005), 499–499.

[26] Snidaro, L. and Foresti, G.L., *Knowledge representation for ambient security*, Expert Systems, **24**(5) (2007), 321–333.

[27] Solberg, A.H.S., Storvik, G., Solberg, R. and Volden, E., *Automatic detection of oil spills in ERS SAR images*, IEEE Transactions on Geoscience and Remote Sensing, **37** (4) (1999), 1916–1924.

[28] Vethamony, P., Sudheesh, K., Babu, M.T., Jayakumar, S., et al., *Trajectory of an oil spill off Goa, eastern Arabian Sea: Field observations and simulations*, Environmental Pollution (2007).

[29] Wooldridge, M. and Jennings, N.R., *Intelligent Agents: Theory and Practice*, The Knowledge Engineering Review, **10**(2) (1995), 115–152.

[30] Yang, J. and Luo, Z., *Coalition formation mechanism in multi-agent systems based on genetic algorithms*, Applied Soft Computing Journal, **7**(2) (2007), 561–568.

Juan Manuel Corchado
Department of Computing Science and Automatic
Plaza de la Merced, s/n
Salamanca
Spain
e-mail: **corchado@usal.es**

Aitor Mata
Department of Computing Science and Automatic
Plaza de la Merced, s/n
Salamanca
Spain
e-mail: **aitor@usal.es**

Sara Rodríguez
Department of Computing Science and Automatic
Plaza de la Merced, s/n
Salamanca
Spain
e-mail: **srg@usal.es**

Whitestein Series in Software Agent Technologies and Autonomic Computing, 119–137

A Methodology for Developing Environmental Information Systems with Software Agents

Ioannis N. Athanasiadis and Pericles A. Mitkas

Abstract. This article presents a unifying methodology for developing environmental information systems with software agents. Based on the experience reported in recent literature, we abstract common requirements of environmental information systems into agent types, combine state-of-the-art tools from computer science, service-oriented software engineering and artificial intelligence domains, as software agents and machine learning, and illustrate their potential for solving real-world problems. Specifically, two generic agent types are specified that behave as information carriers and decision makers, which provide an appropriate abstraction for deployment of added-value services in environmental information systems.

A concrete pathway for applying these instruments throughout the software life cycle of an environmental information system is outlined, along with suggestions for software specification and deployment tools. The method is demonstrated in two application domains: one for air quality assessment and another for meteorological radar data surveillance.

Keywords. Intelligent Information Systems, Agent-oriented Software Engineering, Environmental Data Management, Integration and Reporting, Methodological Tools, Agent Architectures.

1. Introduction

1.1. Environmental Information Systems: Scope and Challenges

Environmental Information System (EIS) is a broad term used for a range of IT systems related to natural resources data management. A working definition, given in [4], is the following one: *"An Environmental Management Information System can be considered as an enterprise information system that provides efficient and accurate access to knowledge elements related to information about the natural environment."*

Environmental monitoring networks established worldwide, primarily in areas with potential pollution problems, observe and record the conditions of the natural environment. Through these networks, vast volumes of raw data are captured, and EISs are in charge of integrating all recorded data-streams. A typical EIS installation involves the fusion into a central database of environmental data recorded at distributed locations and in different means. Most commonly, EISs have been developed and installed to pursue one or more of the following goals:

a. *Off-line analysis systems.* Such systems are geared towards gathering historical data in a systematic way and making them available for in-depth analysis of natural phenomena.

b. *(Real-time) reporting systems.* These are systems responsible for identifying and reporting the current environmental conditions. They satisfy the public need for environmental awareness and administrative and industrial needs for prevention measures.

c. *(Early) Warning Systems.* Their goal is to prognosticate future environmental conditions. The need to forecast and forewarn about potential environmental problems is the key for preserving nature and taking precautionary actions.

Until lately, environmental data were meant for environmental scientists occupied with off-line studies and post-processing activities in their effort to understand the natural phenomena involved. However, there has been a transition in this practice: The aftermath of the growing societal interest in the environment and sustainable development was the emerging need for providing environmental information to the public.

Considering the quest for environmental information involving citizens, industry and administration, the challenge for an EIS is to provide with *advanced information services.* EIS' objectives are no longer restricted to integrate and process raw data-measurements. EISs are challenged to fuse information and diffuse knowledge, in forms comprehensible and accessible by everyone.

One challenge for modern EISs is to broaden their scope and embrace new users from the administration, industry, and the society. Modern EIS users have varying interpretations of environmental values, and consequently different needs in terms of detail of information and mediums for communication, but also conflicting interpretations of the data handled by EISs. In spite of their diverse needs, all users agree on the necessity to access trustworthy information *on time.* One of the major challenges for EISs today is to effectively capture, manage and report environmental information at *"near real-time".* Furthermore, modern EISs are called to develop personalized services, and to tackle issues related to data ownership and permissions, spatial and temporal scaling, industrial patent protection, intellectual property rights, and privacy issues.

Another challenge for EISs is that traditionally EISs are developed for certain case studies, therefore the generalization of the approach and the potential reuse of tools is a very seldom situation. This is partially an intrinsic characteristic of environmental systems, as model configuration and adaptation to local conditions is required. Knowledge sharing, in any forms from raw data to sophisticated environmental model implementations, has become an increasingly important aspect of sound environmental management [4].

The above challenges can be met by adopting modular, service-oriented approaches, such as software agent technology, where reusable components can provide with added-value digital services in open environments.

1.2. Software Agent Technology

Agent-oriented software engineering has emerged as a novel paradigm for building software applications. The key abstraction used is that of an agent, as a software entity characterized by autonomy, reactivity, pro-activity, and social ability [12]. Certain types of software agents are able to infer rationally and support the decision making process [11]. Although there are variant definitions of the notion of agency found in literature (see the discussion in [23]), as a working definition we consider a *standard agent* a computational system operating in some environment, capable for sensing its environment and act upon it, in order to fulfill its goals ([22]). Agent-based systems may rely on a single agent, but the advantages of this initiative are revealed in the case of multi-agent systems, which consist of a community of co-operating agents. Several agents, structured in groups, can share perceptions and operate synergistically to achieve common system goals.

In agent-oriented software engineering, an agent is both a metaphor for software design and an abstraction for software development. As a software design metaphor, agents are considered as the building blocks of a system. Agent related technologies for software design include techniques for system requirements specification, software modeling, specification and verification (see discussion in [21]). Taking a step ahead, agent technology has moved to agent-oriented software engineering that adopts agents in the software design process, as for example in GAIA [26]. For agent-based development, there is a plethora of agent deployment strategies and toolkits (for an extended list the reader is directed to [13]), that vary from object-oriented programming and custom multi-agent systems to service oriented systems and agent platforms [14]. The latter have emerged as the evolution of object-oriented programming and distributed computing, and utilize agents as the basic software unit for developing software.

Agents are well-suited in open, competitive environments, as those of knowledge brokering, personal assistants, and online auctions, just to name a few. According to Parunak [17], software agents are best suited for applications that are modular, de-centralized, changeable, ill-structured, and complex. Parunak draws this conclusion, by ascertaining industrial and commercial applications, mainly in the fields of control systems, enterprise resource planning systems and electronic services.

2. Related Work

EISs bear similar properties with the systems reviewed by Parunak (in [17]): EISs need to address several users at different service levels, integrate data and information from heterogeneous sources, deal with data at multiple spatial and temporal scales and adapt to changing conditions. Also, they inherit both the uncertainty and the complexity involved in the natural phenomena. EISs involve uncertainties both at data, model and decision-making levels, and complexities related to the conflicting requirements and values of the involved users and stakeholders. Consequently, one could claim that the area of environmental informatics fits well with the competencies of agent-based systems.

Agent technology has attracted a significant amount of attention from researchers in environmental informatics. Agent-based approaches have been adopted for developing environmental systems for data management, decision support or simulation purposes. Agent capabilities for distributed problem solving, adaptive personalized services, knowledge sharing and proactive autonomy may enable advanced digital services for EISs.

Since the 1990s, agent technologies have been welcomed in ecological and environmental applications mainly as a metaphor for decomposing complex systems and studying the emergence of collective behavior [16]. Ever since, agent-based techniques have been extensively used for modeling in several environmental fields, including population dynamics, landscape modeling, water management, forest fires simulations, to name a few. However, agent technology has been adopted as a tool for software design and implementation of environmental applications in a limited, rather fragmented way [1]. In a review of agent-based systems applied in environmental informatics [1], Athanasiadis studied twenty three systems that utilize agent technology at different stages of EIS development. An outlook of agent use in environmental software is illustrated graphically in Figure 1 (from [1]). The penetration of agent-oriented tools for software design and development is qualitatively represented on the two axes, and the acronyms of the systems reviewed is situated in the hyperplane.

While there are several systems that use agent modeling for system design, most of them have been implemented using traditional, object-oriented techniques. Implementations that employ agent platforms are still in infancy. Similar holds for agent oriented specification methodologies and tools that employ more sophisticated agent-oriented design toolkits. Only the PICO project [18] reports an agent-oriented software engineering technique throughout the whole design process, while software development using agent-based programming techniques is not accompanied with agent-based design to a great extend.

There is a lot of space for exploiting agent technology in EISs, by adopting agent-oriented software engineering and agent programming techniques in future developments. Also, one of the issues that have not been tackled so far, is a unifying methodology that abstracts common requirements of environmental information

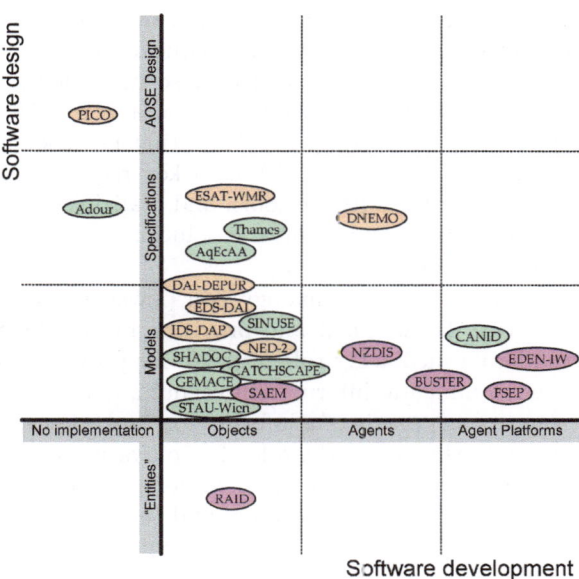

FIGURE 1. Penetration of agent-based techniques for software design and implementation in environmental informatics (from [1]).

systems into agent types and provide an appropriate framework for deploying agent-based EISs.

3. Structural Components of the Methodology

3.1. Agent Orientation in EIS Life Cycle

Advancing on the way earlier research work has dealt with EISs using agent technology, we propose a methodology for developing EISs as multi-agent systems (MASs). Our goal is to assign all services involved in the operational EIS to a software agent society. This work considers the notion of an **agent** as the basic building block for the requirements analysis, software specification, design and development for an EIS.

Agents are treated as *the* common conceptual tool that goes through the whole software development life cycle of an EIS from its conception to final delivery and installation. Given the human-like characteristics of an agent, as their abilities to shape behaviors, realize roles and establish communicative dialogues, it makes them a very handy tool for modular, decentralized, integrated software engineering process of an EIS. Agent orientation in EIS development can assist with tackling problems involved in environmental applications as domain complexity and interdisciplinarity. An agent in this way can be seen as a useful metaphor that is much easier for wider audiences to comprehend, in contrast with the conventional software engineering paradigms. As a consequence, environmental scientists or the final users are enabled to follow easily the all software development phases of an EIS. Transparency in environmental software and accessibility to an extended peer community, involving other computer practitioners, disciplinary scientists, users and affected public, is considered [19] as a key requirement for increasing the trust against environmental software tools, and realizing the uncertainties involved. Agent orientation can provide natural solutions towards the direction of transparent, modular solutions.

Furthermore, software agent technology is a powerful tool for the development of advanced service development and provision in an EIS. Software agents are capable to formulate a mediating role, capable for providing services in open environments. The environmental information vacuum, underlined by Agenda 21, could be bridged using agents playing an information brokering role among diverse stakeholders and users. In this sense, the role of a software agent in environmental informatics is anticipated from a service oriented perspective. Software agents are service providers, mediating between end users and the environmental data pools for providing advanced information and decision support services.

We identify two main functions of software agents in EISs: software agents that behave as **information-carriers** or **decision-makers**. Agents as information carriers, act as a distributed community of data processing units, able to capture, manipulate and propagate information efficiently, i.e. they provide with data manipulation services. Agents as decision-makers, behave as a network of problem-solvers that work together to reach solutions.

In the followings we specify the behavior of an agent in EISs by defining a common external view, that determines how agents are interacting with their virtual environment, and two internal views: one for the information-carrier type and one for the decision-maker type. The internal views set out agent's own private behavior and functionality.

3.2. An Abstract Agent for EISs (External View)

Based on the mediating profile of a software agent in environmental informatics, and using the notion of a standard agent [22], we define at an abstract level, an agent (*agent*) in EISs as an autonomous entity that defines its actions based on its own perceptions about the state of its (virtual/artificial) environment.

Information on the state and the conditions of natural environment is captured in the form of **environmental data**. Environmental data may be a result of inspection, measurement, or simulations, and typically have spatio-temporal references.

Definition 3.1. Environmental data objects (EDO) constitute the virtual environment in which an *agent* operates. The (virtual/artificial) environment of an agent comprise all the possible states of the environmental data objects it percepts.

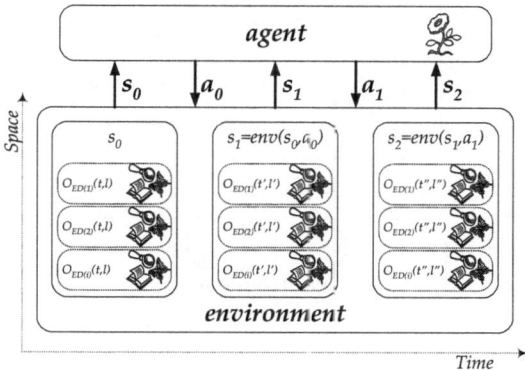

FIGURE 2. Software agent - virtual environment interaction.

Let $\mathbf{O}$ be the set of environmental data objects that *agent* percepts. In an EIS each EDO can be considered as a function of space s and time t, therefore *agent* may be potentially exposed to a set of EDOs:

$$\mathbf{O} = \{O_{ED(1)}, O_{ED(2)}, \dots, O_{ED(i)}, \dots\} \qquad \text{where } O_{ED(i)} = f(s, t).$$

The set $\mathbf{S}$ of the environmental states that an agent may percept is defined as the set of all instances of its environment. In principle, the states to which an agent is exposed to is an infinite set:

$$\mathbf{S} = \{s_1, s_2, \dots\}, \forall s_j = O_{ED(i)}(s, t).$$

Definition 3.2. Let $\mathbf{A} = \{a_1, a_2, \dots\}$ be the set of the possible actions of an *agent*, then the *agent* can be defined as a function that maps the sequences of the environmental states to agent actions as:

$$action : \mathbf{S}^\star \to \mathbf{A}. \tag{3.1}$$

While the environment of the *agent* reacts to the action $a \in \mathbf{A}$ applied on state $s \in \mathbf{S}$ as:

$$env : \mathbf{S} \times \mathbf{A} \to \mathbf{S}. \tag{3.2}$$

Definition 3.3. The execution (*run*) of *agent* is the sequence:

$$run : s_0 \xrightarrow{a_0} s_1 \xrightarrow{a_1} s_2 \xrightarrow{a_2} s_3 \dots \tag{3.3}$$

where $s_1 = env(s_0, a_0)$ is the state in which the agent environment goes when the action a_0 is performed on state s_0. The interaction between an agent and its environment is illustrated in Figure 2, that defines the external view of an *agent* of the toolbox.

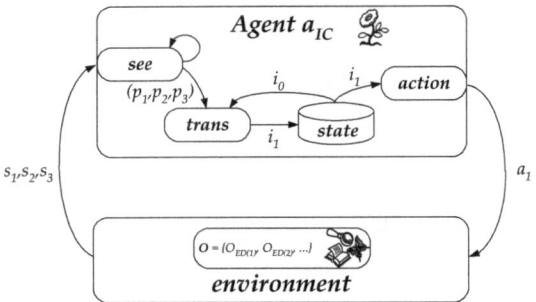

FIGURE 3. Internal structure and behavior of agent a_{IC}.

Having defined the external view of a generic agent in an EIS, we proceed with the specification of the internal views of the two abstract agent types: the information carrier agent and the decision-maker agent.

3.3. Information-carrier Agent Internal View

The information-carrier agent role a_{IC} operates as a function that transforms EDO, in order to provide added-value data transformation and dissemination services. From an external point of view, agent a_{IC} percepts a series of environmental states $\mathbf{S} = \{s_1, s_2, s_3, \dots\}$ to which it responds with a series of agent actions $\mathbf{A} = \{a_1, a_2, a_3, \dots, a_i\}$. Each action $a_i \in \mathbf{O}^\star$ i is an environmental data object that alters its environment's state.

Definition 3.4. The information carrier agent a_{IC} is an agent with state. The internal view on a_{IC} agent behavior is illustrated in Figure 3. Each internal state i_k is a set of EDO instances, therefore:

$$\mathbf{I} = \{i_1, i_2, i_3, \dots i_k\}, \qquad i_k \in \mathcal{P}(\mathbf{O}). \tag{3.4}$$

The information carrier agent a_{IC} operates as follows: It observes its virtual environment through a perception function *see*, it captures the environmental states $s \in \mathbf{S}$ into agent perceptions $p \in \mathbf{P}$. Based on the sequences of perceptions, agent a_{IC} refreshes its internal state through the transformation function *trans*. Finally, a_{IC} performs its actions, based on its internal states via the *action* function.

Definition 3.5. The a_{IC} "internal state" can be specified as:

$$see \quad : \quad \mathbf{S} \to \mathbf{P} \tag{3.5}$$
$$trans \quad : \quad \mathbf{I} \times \mathbf{P}^\star \to \mathbf{I} \tag{3.6}$$
$$action \quad : \quad \mathbf{I} \to \mathbf{A} \tag{3.7}$$
$$\mathbf{P}, \mathbf{S} \subseteq \mathbf{O}, \mathbf{I} \subseteq \mathcal{P}(\mathbf{O}), \text{ and } \mathbf{A} \subseteq \mathbf{O}^\star$$

Implementing this behavior an information carrier agent is capable of managing EDOs as follows: Suppose that initially a_{IC} enjoys the internal state i_0 and at some point it is exposed to three environmental states: s_1, s_2, s_3. Through the function *see* these environmental conditions are perceived by the a_{IC} as $p_1 = see(s_1), p_2 = see(s_2), p_3 = see(s_3)$. Then, the transformation function *trans* drives agent a_{IC} to the internal state $i_1 = trans((p_1, p_2, p_3), i_0)$, which causes it to return to its environment an action (sequence of EDOs) $a_1 = action(i(1)) = action(trans((p_1, p_2, p_3), i_0)$. This behavior is illustrated in Figure 3.

The abstract behavior specified, enables a_{IC} to perform a variety of data manipulation activities. From simple data capturing activities (i.e. from sensor networks) and database queries, to data aggregation and scaling, as well as complex transactions in an information system.

The set of all information carrier agents is noted by IC and contains all agents that implement the a_{IC} behavior.

3.4. Decision-maker Agent Internal View

An agent a_{DM} functions as a decision maker by incorporating a reasoning engine *engine*, that implements a decision-making model. The decision making model can encompass deterministic strategies, knowledge-discovery techniques or heuristics.

Definition 3.6. The reasoning engine of a decision-maker agent a_{DM} is a mapping of internal states $i_1, i_2, \ldots i_n$ of a_{DM} to a decision d, following the relation:

$$engine : \mathbf{I}^\star \to \mathbf{D}. \tag{3.8}$$

Agent a_{DM}, based on its decisions $d \in \mathbf{D}$ responds to the stimuli of its environment by performing a set of actions $c \in \mathbf{A}$.

Definition 3.7. Following the generic model of agent with state, agent a_{DM} operates as follows:

$$
\begin{aligned}
see &: \mathbf{S} \to \mathbf{P} & (3.9)\\
next &: \mathbf{I} \times \mathbf{P} \to \mathbf{I} & (3.10)\\
engine &: \mathbf{I}^\star \to \mathbf{D} & (3.11)\\
action &: \mathbf{D} \to \mathbf{A} & (3.12)
\end{aligned}
$$

$$\mathbf{P}, \mathbf{S} \subseteq \mathbf{O}, \mathbf{I} \subseteq \mathcal{P}(\mathbf{O}), \text{ and } \mathbf{D} \subseteq \mathbf{OA} \subseteq \mathbf{O}^\star$$

Let a_{DM} be in state i_0 and that at a certain point it is stimulated by observing an environmental state s_1. Through its *see* function it shapes the perception $p_1 = see(s_1)$, and consequently through function *next*, a_{DM} revises its internal state to $i_1 = next(i_0, p_1)$. Suppose that the state sequence (i_1, i_2) results the reasoning engine *engine* to get the decision $d_1 = engine(i_1, i_2)$. Due to this decision, a_{DM} performs the action $a_1 = action(d_1) = action(engine(i_1, i_2))$. The internal state and the behavior of an decision maker agent role are illustrated in Figure 4.

With the above internal model, we specify an agent with ability to infer, based on its virtual environment observations, to shape its perceptions and revise

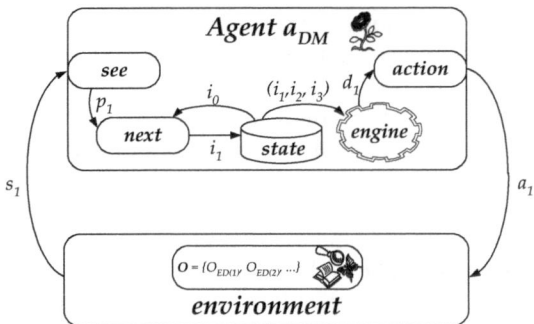

FIGURE 4. Internal structure and behavior of agent a_{DM}.

its internal state. Based on the sequence of its internal states it makes its decisions that feedbacks as actions to its environment. The set of all decision-maker agents is noted as $\mathcal{DM}$, that contains all agents that implement the behavior of $a_{DM} \in \mathcal{DM}$.

4. Deployment of the Toolbox

4.1. Multi-agent Systems for EIS Design and Deployment

The advantages of agent technologies are revealed in the case of multi-agent systems, which consist of a community of co-operating agents. For the design and the development of EISs more than one agent of the two types defined above are cooperating together.

Definition 4.1. A multi-agent environmental information system model ($EnvMAS$) can be specified as a set of N agent types ag_n, each one of which implements the behavior of the information carrier agent type, or the decision-maker type.

$$EnvMAS_{model} = \{ag_1, ag_2, ag_3, \ldots, ag_n, \ldots, ag_N\} \qquad (4.1)$$
$$ag_n \in \{\mathcal{IC} \cup \mathcal{DM}\}, \qquad n = 1 \ldots N.$$

Each agent type ag_n may have one or more instantiations $ag_n(k)$, $k = 1 \ldots K_n$, and therefore the operating multi-agent system is specified as:

$$
\begin{aligned}
EnvMAS \;=\; & \{ag_1(1), ag_1(2), \ldots ag_1(K_1), \qquad (4.2)\\
& ag_2(1), ag_2(2), \ldots ag_2(K_2),\\
& \ldots, ag_n(1), ag_n(2), \ldots, ag_n(K_n),\\
& \ldots, ag_N(1), ag_N(2), \ldots, ag_N(K_N)\}.
\end{aligned}
$$

In total, there are $\sum_{n=1}^{N} K_n$ agents running in the system. $EnvMAS$ interacts with the natural environment and the end users through EDOs perceived by the agents, while inter-agent communication ensures system coordination.

4.2. How to Get the Toolbox to Work

Having defined a generic agent for EISs, two specific agent types and a multi-agent system in our toolbox, the question that rises is how to use them in order to design and deploy an EIS as a multi-agent system. Though there are no silver bullets in software design, here we present a pathway that starts from an (unknown) application domain, decomposes it into agents of the toolbox, and synthesize them into an operating multi-agent system. An agent is a metaphor that is used throughout the software life cycle. The specified abstract agent types are to be used for system analysis, software design and specification and software development.

4.2.1. System Analysis Phase. As in any software project, first comes the *system analysis phase*, which starts with a study of the domain at hand. Problem definition and the specification of the system goals is performed using abstract agent types as roles for defining system components and functions. This step can be realized by specifying the entities of the system using the toolbox. In principal, we identify two categories of entities:

(a) Entities that are part of the problem and contribute in the system specification.
(b) Entities that are part of the system and frame the problem boundaries.

In the first group fall all system drivers, which are external to the software system, as the end users. Such entities influence system specification, but they are not part of it. The second category comprise those entities which are part of the system. This role can be played by humans (i.e the system administrator), parts of the natural environment (i.e a river), or hardware or software components of the system (i.e a sensor network or a database system).

The analysis phase concludes with the specification of the entities, their behavior and their interaction. This process can be done using requirement elicitation methods, as role playing games, where all stakeholders are involved. Environmental scientists, decision-makers, software architects and end users are engaged in this phase.

4.2.2. System Design Phase. Next comes the *system design phase*, which involves four main steps:

1. Entity behavior is assigned to software agent types.
2. The general system architecture is defined.
3. System functionalities are specified.
4. The system is specified using agent-based models.

The first step is to match system entities to certain agent types. The criterion for the assignment is the functionality and the behavior of entities. Information processing functionalities are assigned to information carrier agent types, and decision making nodes to decision maker agents. This step sketches a first draft of the system design, based on the requirements defined during the analysis phase.

Next, the general system architecture is specified, based on the envisioned services functionality. Agent behaviors are interwoven to ensure certain information flows through agent cooperation and coordination. Protocols for agent communication are identified and the external views of the agents are specified.

The third step is the functional specification of the system. Each of the agents is specified in detail and the internal views are detailed. Specifically, agent transformation functions are defined for information carrier agents. For decision-maker agents, internal states and inference engines are specified. Agent design is further discussed below in paragraphs 4.2.3 and 4.2.4.

The design phase is concluded with the agent-oriented system design phase. Agent modeling toolkits, as GAIA [24, 26], are used for the detailed system specification, while the in-depth agent communication can be designed with AORML [20]. The above mentioned tools are a suggestion for specifying a multi-agent system as a computational organization; software architects may select alternatives for agent-oriented system design, as AUML [15], iSTAR [25], or Tropos [10].

4.2.3. Information Carrier Agent Design. Modeling agents as information carriers involves four steps:

Step 1. *Identify system inputs and outputs*: Consider the interfaces between the software system, data sources and end-users. Identify services provided by system entities. Assign agents to realize these interfaces acting either as data fountains, or data sinks.

Step 2. *Formulate information channels*: Detail how information flows through the system. Specify possible data transformations needed. Assign those tasks to information carrier agents that operate as data managers, across data fountains and sinks.

Step 3. *Conceptualize agent messaging*: Based on the two previous steps, realize inter-agent communications for smooth information propagation. Specify the semantics of the communications using ontologies.

Step 4. *Specify delivery deadlines*: Concrete deadlines are assigned to agent communication, in order to ensure 'on-time' delivery of information. Exit on failure strategies need to be detailed too.

The outcome of the above procedure is materialized as the specification of a MAS architecture in the form of:

$$\text{MAS} = \langle A, O, I, D \rangle \tag{4.3}$$

where:

- $A = \{\text{ag}_1, \ldots, \text{ag}_n\}$, is a countable set of software agents.
- O is the domain ontology, which specifies the common vocabulary in order to represent the system environment.
- $I = \{\text{I}_k = (\text{ag}_i, \text{ag}_j)/\text{ag}_i, \text{ag}_j \in A\}$, is a set of interactions between agents. These interactions show the relations in the system organization and they allow the definition of a social framework determining the information flows in the system.
- $D = \{\text{D}_k, \forall \, \text{I}_k \in I\}$, is a set of the delivery deadlines assigned to each agent communication.

4.2.4. Decision-maker Agent Design. Agents as decision makers are employed to deliver the reasoning abilities of the EIS. Indicatively, decision-making in an EIS may involve assessment services or activities to overcome data uncertainty problems. Based on the domain knowledge, agent decision-making strategies are identified through the following procedure:

Step 1. *Problem formulation and decomposition*: Consider the overall problem at hand and try to break it down into sub-problems.

Step 2. *Construction of decision points*: Assign specific agents to solve each sub-problem, taking under account their resources, specified by the system's architecture.

Step 3. *Decision strategy specification*: For each sub-problem provide a strategy to solve it using the available resources.

Step 4. *Realization of Inference models*: Implement the decision strategies designed in the previous step as inference models of the respective agents. Inference models will be embedded into decision-maker agents as reasoning engines.

This procedure is highly dependent on the application under consideration. Finding an optimal decision strategy is a rather difficult task, especially when execution time is a parameter of success. However, three distinct cases of decision-making engines, can be identified, covering the majority of applications:

Case 1. *Deterministic Strategies*: These are applied, when domain-specific, certain, explainable rules for decision-making are available. Such rules may encompass natural laws, logical rules or legal constraints. In such cases, rules are incorporated as a static, confident, explainable expert system into the agents.

Case 2. *Data-driven Strategies*: When historical datasets are available, the application of machine learning algorithms for knowledge discovery can yield interesting knowledge models. These models can be used for agent reasoning in a dynamic, inductive way. In EISs, there are large volumes of data continually recorded. When natural laws describing the monitored phenomena do not exist, or they are too complex, data-driven models, such as decision trees, case-based reasoning, or neural networks is an alternative to the system architect. In this case, the procedure involves the creation of an inference model from historical data. This model is later incorporated into the agents.

Case 3. *Heuristic strategies*: When neither of the above cases is applicable, heuristic models or 'rules of thumb' may be adopted for agent reasoning.

This checklist provides with a guideline for designing decision-making agents required by a multi-agent EIS.

4.2.5. System Development and Deployment Phase. *System development* is the third phase of the process. Having specified the overall agent architecture, the internal agent structures and agent communication protocols in the previous step, next comes the system deployment using an agent platforms. Software engineer has a plethora of tools available for agent programming and deployment, as JADE [9, 8]. Agent programming toolkits consist a middleware for the development of

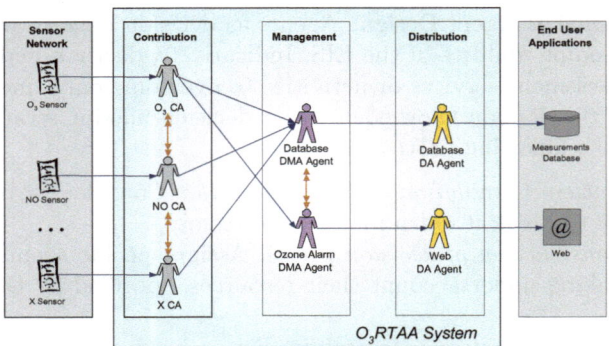

FIGURE 5. O₃RTAA System Architecture.

distributed multi-agent applications, that support natively peer-to-peer agent communication, basic agent behaviors and an agent runtime environment.

The development phase concludes with system installation and deployment. An iterative process for revealing design faults or development bugs is then required for ensuring software quality of the system's final version.

5. Demonstration of the Methodology

The methodology presented here has been demonstrated in two case studies which we discuss below.

5.1. Application in Air Quality Assessment

The methodology described in the previous section has been applied to the development of O₃RTAA, an EIS for air quality assessment and reporting. O₃RTAA is a multi-agent system for monitoring and assessing air quality, by exploiting data from a sensor network. A community of software agents is assigned to monitor and validate measurements coming from several sensors, to assess air-quality, and, finally, to fire alarms to appropriate recipients, when needed, via the Internet. The overall system architecture is depicted in 5.

In O₃RTAA, information carrier agents are responsible to collect data from field sensors, perform data management activities, as data preprocessing, normalization and transformation, and propagate information, which involves posting information to end-users over the internet and updating a measurement database. As shown in Figure 5, "contribution agents" (CA) operate as data fountains of the system, which capture data from the sensors and "distribution agents" (DA) are data sinks which provide with information services to end users.

Decision-maker agents in O₃RTAA are responsible for validating incoming measurements; substituting erroneous measurements by estimating missing values and approximating false sensor readings; and calculating of qualitative indicators.

The first two activities are left to CA agents, while "Alarm DMA" agent is in charge of the third. Data-driven strategies are employed for data validation and erroneous data substitution engines, while deterministic strategies were used for air quality indicator engine.

The O$_3$RTAA system has been tested against real data and demonstrated as a pilot case at the Mediterranean Centre for Environmental Studies Foundation (CEAM), in collaboration with IDI-EIKON, Valencia, Spain. More details on O$_3$RTAA architecture are given in [6] and a more generalized framework is presented in [2]. Data-driven strategies using knowledge discovery techniques are presented in [7, 3]. O$_3$RTAA has demonstrated an open architecture for fusing sensor data related to air quality, exploited data-driven techniques for quality assurance and incorporated legal restrictions for issuing air quality warnings. The benefits of the approach include system extensibility, as new agents may be added at runtime for providing with new services; and system modularity as data quality assurance problems are tackled step-by-step and diverse user types are treated by different agents.

5.2. Application in Meteorological Radar Data Surveillance

The second case study is a meteorological radar data surveillance system deployed as a pilot service for the Meteorological Service of Cyprus. The agent-based EIS developed, called ABACUS intervenes between a meteorological Doppler radar and end-users, as the meteorological service and the local airport. The goal of the system is to manage and process radar recordings (which indicate clouds formations above the island); identify specific meteorological incidents and their evolution through time, and to provide with digital services to the end users, as online warnings and visualizations.

This goal was assigned to a community of cooperating agents, illustrated in Figure 6. An information carrier agent is responsible for acquiring radar scans and preprocess them by applying certain filters, acting as the system's data fountain. A set of meteorologist agents (decision-maker agents), each one of which is responsible for an annular sector within the radar's range, calculates metrics and indices within its sector and applies decision rules for assessing the weather conditions and issuing alarms. Meteorologist agents incorporate deterministic and heuristic strategies for assessing meteorological conditions. Deterministic strategies have to do with verified patterns, while heuristics incorporate expert knowledge. Finally, a couple of information carrier agents further process the data, persist them in a database, render graphs, maps and other visualizations and present them to the end users.

ABACUS has been demonstrated with real data at the Meteorological Service of Cyprus. System architecture is detailed in [5]. ABACUS has demonstrated how a software agent society may assist with a laborious task, by undertaking some of the human experts' tasks, while adopts their own criteria. Runtime customization and user-tailored system behavior, along with the automation of radar data filtering and visualization are the strong points of the system.

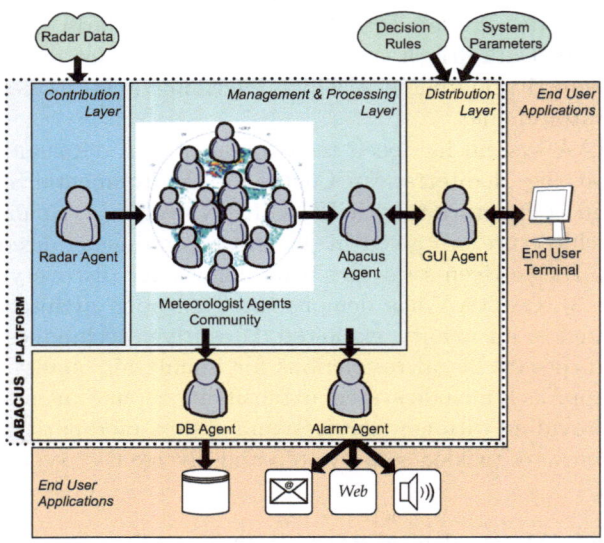

FIGURE 6. Abacus system architecture.

6. Discussion

In this work, we presented an methodology for developing agent-based EIS, that supply with digital services, as those of seamless data integration, environmental assessment, warning services, information diffusion, and advanced decision making. The method relies on two generic agent types for EISs and a concrete pathway for using agents as a unique metaphor for system analysis, design and deployment.

An overview of the methodology is depicted in Figure 7. It unifies in a single process two properties of agents: their capacity (a) for distributed information processing, and (b) for distributed problem solving. The main advantage of such an approach is that it fully exploits the capabilities of autonomous agents, considering them as both information carriers and decision makers. Information flows dictate how agents manipulate data, while domain knowledge determines the decision making process incorporated into the agents. Information flows are implemented through agent communication channels, while the decision-making processes are transformed into agent reasoning models. Agent architectures that can be designed with our methodology are able to deal with data uncertainty problems, through the hybrid use of either deterministic, data-driven or heuristic decision-making strategies for agent reasoning.

Our methodology provides with the means for adopting agent technology throughout the life cycle of environmental information systems. We defined where and how agents can be used, defined abstract agent types, suggested existing tools

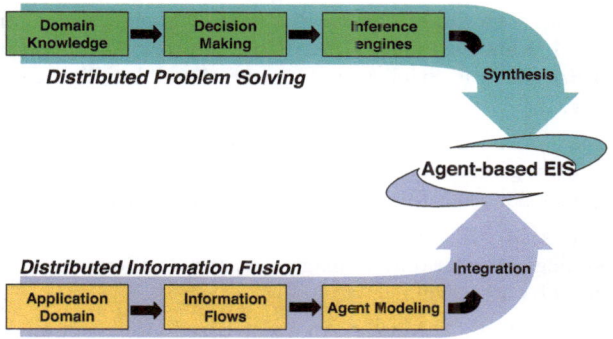

FIGURE 7. An abstract view of our methodology.

from agent oriented software engineering and provided with guidelines about the process that a software architect has to follow for developing an agent-based EIS.

The benefits of our methodology rely on two pillars: First is the use of agents for software requirement analysis and design, as the human like characteristics of agents are much easier for the environmental scientist to comprehend and communicate with. This enables the environmental scientists can be engaged deeper in the EIS development process. Second, it employs a distributed information processing approach, using software agents, thus agent-based EISs are open, modular and extensible, which is always a goal for EISs. Finally, we argued that using a single metaphor (that of an agent) throughout the EIS life cycle is a great advantage for building trust of environmental scientists in EISs.

The article demonstrated how agent technology can be applied for meeting the challenges of developing modern EISs. A concrete methodology for the software developer has been proposed and demonstrated in two applications. Future efforts will concentrate in extending the tools presented for agent-based social simulations, which is a very popular domain in environmental modeling. Follow up activities are required for studying issues related to the semantics of environmental data brokering, focusing on problems related to scaling, privacy and ownership. Sophisticated agent behaviors may be the key for addressing such challenges, in distributed, open environments.

References

[1] I. N. Athanasiadis. A review of agent-based systems applied in environmental informatics. In A. Zerger and R. M. Argent, editors, *MODSIM 2005 Int'l Congress on Modelling and Simulation*, pages 1574–1580, Melbourne, Australia, December 2005. Modelling and Simulation Society of Australia and New Zealand.

[2] I. N. Athanasiadis. An intelligent service layer upgrades environmental information management. *IT Professional*, 8(3):34–39, May-June 2006.

[3] I. N. Athanasiadis. The Fuzzy Lattice Reasoning Classifier for mining environmental data. In V. G. Kaburlasos and G. X. Ritter, editors, *Computational Intelligence Based on Lattice Theory*, Studies in Computational Intelligence, pages 175–193. Springer-Verlag, 2007.

[4] I. N. Athanasiadis. Towards a virtual enterprise architecture for the environmental sector. In N. Protogeros, editor, *Agent and Web Service Technologies in Virtual Enterprises*, pages 256–266. Information Science Reference, 2007.

[5] I. N. Athanasiadis, M. Milis, P. A. Mitkas, and S. C. Michaelides. A multi-agent system for meteorological radar data management and decision support. Environmental Modelling and Software, 2009.

[6] I. N. Athanasiadis and P. A. Mitkas. An agent-based intelligent environmental monitoring system. *Management of Environmental Quality*, 15(3):238–249, 2004.

[7] I. N. Athanasiadis and P. A. Mitkas. Knowledge discovery for operational decision support in air quality management. *Journal of Environmental Informatics*, 9(2):100–107, Jul 2007.

[8] F. Bellifemine, G. Caire, A. Poggi, and G. Rimassa. JADE-A white paper. *EXP in search of innovation*, 3(3):6–19, September 2003.

[9] F. Bellifemine, A. Poggi, and G. Rimassa. Developing multi-agent systems with JADE. In *Proc. of 7th Int'l Workshop on Agent Theories, Architectures and Languages (ATAL-2000)*, Boston, MA, 2000. Available online:http://jade.cselt.it.

[10] F. Giunchiglia, J. Mylopoulos, and A. Perini. The Tropos software development methodology: processes, models and diagrams. In F. Giunchiglia, J. Odell, and G. Weiss, editors, *Software Engineering III, Third International Workshop, AOSE-2002*, LNCS. Springer-Verlag, 2003.

[11] N. Jennings, K. Sycara, and M. J. Wooldridge. A roadmap of agent research and development. *Autonomous Agents and Multi-Agent Systems*, 1(1):7–38, 1998.

[12] N. R. Jennings. On agent-based software engineering. *Artificial Intelligence*, 117:277–296, 2000.

[13] M. Luck, P. McBurney, and C. Preist, editors. *Agent Technology: Enabling Next Generation Computing, A Roadmap for Agent Based Computing*. AgentLink, 2003.

[14] E. Mangina. *Review of software products for Multi-Agent Systems*. AgentLink, 2002.

[15] J. Odell, H. v. D. Parunak, and B. Bauer. Extending UML for agents. In *Proc. of the 2nd Int'l Workshop on Agent-Oriented Information Systems*, Berlin, Germany, 2000. iCue Publishing.

[16] R. L. Olson and R. A. Sequeira. An emergent computational approach to the study of ecosystem dynamics. *Ecological Modelling*, 79:95–120, 1995.

[17] H. v. D. Parunak. Agents in Overalls: Experiences and Issues in the Development and Deployment of Industrial Agent-Based Systems. *International Journal of Cooperative Information Systems*, 9:209–227, 2000.

[18] A. Perini and A. Susi. Developing a decision support system for integrated production in agriculture. *Environmental Modelling & Software*, 19:821–829, 2004.

[19] J. Rotmans. Methods for Integrated Assessment: The challenges and opportunities ahead. *Environmental Modeling and Assessment*, 3:155–179, 1998.

[20] G. Wagner. The Agent-Object-Relationship metamodel: Towards a unified conceptual view of state and behavior. *Information Systems*, 28(5):475–504, 2003.

[21] G. Weiss. Agent orientation in software engineering. *Knowledge Engineering Review*,
16(4):349–373, 2002.

[22] M. Wooldridge. Intelligent Agents. In G. Weiss, editor, *Multiagent Systems: A modern approach to distributed Artificial Intelligence*, chapter 1, pages 27–78. MIT Press,
2000.

[23] M. Wooldridge and N. R. Jennings. Intelligent Agents: Theory and Practice. *Knowledge Engineering Review*, 10(2):115–152, 1995.

[24] M. Wooldridge, N. R. Jennings, and D. Kinny. The GAIA Methodology for
Agent-Oriented Analysis and Design. *Autonomous Agents and Multi-Agent Systems*,
3(3):285–312, 2000.

[25] E. Yu. Towards Modelling and Reasoning Support for Early-Phase Requirements
Engineering. In *Proc. of the 3rd IEEE Int. Symp. on Requirements Engineering*,
Washington, USA, 1997. IEEE.

[26] F. Zambonelli, N. R. Jennings, and M. Wooldridge. Developing multiagent systems:
the GAIA Methodology. *ACM Trans on Software Engineering and Methodology*,
12(3):317–370, 2003.

Ioannis N. Athanasiadis
Dalle Molle Institute for Artificial Intelligence,
Lugano, Switzerland
e-mail: `ioannis@athanasiadis.info`

Pericles A. Mitkas
Electrical and Computer Engineering Dept, Aristotle University of Thessaloniki,
Thessaloniki, Greece
e-mail: `mitkas@eng.auth.gr`

Whitestein Series in Software Agent Technologies and Autonomic Computing, 139–171
© 2009 Birkhäuser Verlag Basel/Switzerland

Cognitive Agents as a Design Metaphor in Environmental-Knowledge Management

Luigi Ceccaroni, Alfredo Simón-Cuevas, Alejandro
Rosete-Suárez and Mailyn Moreno-Espino

To all the kids who will have to clean up our mess in the future.

Abstract. Representing and sharing knowledge has been a central problem
in artificial intelligence since its inception. Representations such as semantic
networks, frames, concept maps and ontologies, as well as various methodologies for using these systems have been proposed for dealing with such
issues. However, problems exist about issues such as communication among
heterogeneous agents, incomplete or uncertain knowledge, imprecise formal-
izations, and so on. Here, a mapping system between knowledge representations (concept maps and ontologies) is modeled using a methodology for
the development of multi agent systems. Ontologies are formalized from non-
formal concept maps and can then be used to represent agents knowledge,
and to facilitate the communication among persons and software agents. A
system is presented, in which a set of agents, implementing three functionalities (retrieval, disambiguation and formalization), collaborates in the process
of knowledge management. This multi-agent system is part of a larger knowledge management system based on concept maps, and facilitates the use of
generated and managed knowledge by not only people but also other software
agents, namely those which require knowledge about domains that have been
represented as concept maps, such as the environment domain, object of this
book.

Keywords. Cognitive Agents, Concept Maps, Environmental Knowledge, Ontologies, Semantics.

1. Preamble

Artificial intelligence provides a variety of useful techniques which can be applied
to environmental science to improve knowledge management and problem solving.
In this article we will see a system that integrates two knowledge representations

(described in detail below): human-friendly concept maps and machine-friendly ontologies. This system has to be able to deal with a fair amount of natural-language processing, in particular with concept-sense disambiguation, in a context (environmental issues) in which complexity of representation is normal and unavoidable.

Cognitive agents are proposed as the metaphor for the design of such a system and provide software designers and developers with a way of structuring the application around autonomous, communicative components. Agent-based systems are one of the most vibrant and important areas of research and development to have emerged in information technology in recent years, underpinning many aspects of broader information society technologies. The new metaphor *computation as interaction* is leading to new ways of conceiving, designing, developing and managing computational systems. In this model, applications are societies of components.

These components are viewed as providing services to one another. The components and their services may be owned and managed by different organizations, and thus have access to different information sources and have different objectives. The components are not necessarily activated by human users but may also carry out actions in an automated and coordinated manner when certain conditions hold. These preconditions may themselves be distributed across components, so that action by one component requires prior co-ordination and agreement with other components.

This new metaphor of computing as social activity, as interaction between independent and sometimes intelligent entities, adapting and co-evolving with one another, can be exploited with agent technologies. An agent is a computer program capable of flexible and autonomous action in a dynamic environment, usually an environment containing other agents. In this two-level abstraction, we have encapsulated autonomous and intelligent software entities, called agents, and we have demarcated the society, in which they operate, a multi-agent system.

In the sense that it is a new paradigm, agent-based computing is disruptive. It causes a re-evaluation of the very nature of computing and computational systems, through concepts such as autonomy, coalitions and ecosystems, which make no sense to earlier paradigms. Because of its horizontal nature, it is likely that the successful adoption of agent technology will have a profound, long-term impact on the way in which future computer systems will be conceptualized and implemented.

Agent technologies can be considered from three perspectives: agents as Design Metaphor, as a Source of Technologies, and as Simulation. In the context of the system described in this article, agents are considered as a design metaphor and lead to the construction of components-based software tools and infrastructure. In this sense, they offer a new and often more appropriate route to the development of complex computational systems, especially in open and dynamic environments. In order to support this view of systems development, particular tools and techniques need to be introduced. For example, methodologies to guide analysis and design are required, agent architectures are needed for the design of individual software components, tools and abstractions are required to enable developers to deal with the complexity of implemented systems, and supporting infrastructure (embracing other relevant, widely used technologies) must be integrated [20].

1.1. Technological Context: The Semantic Web

Base technologies of particular relevance include the following:

- The *extensible markup language* (XML) is a language for defining markup languages and syntactic structures for data formats. Though lacking in machine-readable semantics, XML has been used to define higher-level knowledge representations that facilitate semantic annotation of structured documents on the Web.
- The *resource description format* (RDF) is a representation formalism for describing and interchanging metadata.

There are several key *trends and drivers* that suggest that agents and agent technologies will be vital. The work described in this article is part of the current impetus for use and deployment of both agent systems and semantically-rich languages.

Since it was first developed in the early 1990s, the *World Wide Web* has rapidly and dramatically become a critically important and powerful medium for communication, research and commerce. However, the Web was designed for use-by-humans, and its power is limited by the ability of humans to navigate the data of different information sources. A somewhat similar situation is the one of *concept maps* (see below), which were also designed in the 1980s for use-by-humans and are now not compatible with a machine-dominated management of information.

The *semantic Web* is based on the idea that the data on the Web can be defined and linked in such a way that they can be used by machines for the automatic processing and integration across different applications [4]. This is motivated by the fundamental recognition that, in order for Web-based applications to scale, programs must be able to share and process data, particularly when they have been designed independently. The key to achieving this is by augmenting Web pages with descriptions of their content in such a way that it is possible for machines to reason automatically about that content. Among the particular requirements for the realization of the semantic Web vision there are:

- rich descriptions of media and content to improve search and management;
- rich descriptions of Web services to enable and improve discovery and composition;
- common interfaces to simplify integration of disparate systems;
- a common language for the exchange of semantically-rich information between software agents.

It should be clear from this that the semantic Web demands effort and involvement from the fields of agent-based computing and knowledge representation, and that the three fields are intimately connected. Indeed, the semantic Web offers a rich breeding ground for both further fundamental research on ontologies and a whole range of agent applications that can (and should) be built on top of it [20].

1.2. Knowledge Representation in Environmental Sciences

In the environmental sciences, as in most scientific domains, information needs sometimes to be analyzed and processed by machines. In the knowledge representation oriented to the semantic analysis and processing by machines, context in which a certain degree of formalization is required, the development and use of *ontologies* is increasingly common. However, the processes of designing and creating ontologies, the tools available, and the specification languages are still complex for non-experts in this subject. This complexity represents a difficulty in environments requiring the collaboration of humans for the development and processing of ontologies. This suggests that a form of representation that can be used naturally by humans and integrated with ontologies (in such a way that the latter can be automatically obtained) should be useful. Concept maps (see §1.3) are one of these human-friendly knowledge-representations.

1.3. Concept Maps

Concept maps (CMs) are graphical tools for organizing and representing knowledge [26] [25], and were defined for application in the learning process; they are easy to be created, flexible and intuitive for people. They include *concepts*, *linking words or linking phrases* (*lw* or *lp*), to specify the type of relationships between concepts, and *propositions*, which contain two or more concepts, connected using linking-phrases to form a meaningful statement. The concepts and propositions are represented in a hierarchical fashion [7] with the most inclusive, most general concepts (the *root concepts*) at the top of the map and the more specific, less general concepts arranged hierarchically below. Figure 1 shows a CM about the nitrogen cycle, the natural circulation of nitrogen by living organisms via the atmosphere, oceans and soil, and is an enriched interpretation of the representation in Figure 2. CMs are a kind of semantic network, but one that is more flexible and non-formal, oriented to be used and interpreted by humans. CMs propositions can be seen as simplified natural-language phrase; *e.g.* the proposition *(Water, made of, Oxygen)*, in Figure 1, corresponds to the phrase *water is made of oxygen*. This suggested, for analyzing the language represented in the CMs, the usefulness of tools commonly used in natural language processing, such as WordNet [23].

The integration between CMs and ontologies, specifically in the case of OWL ontologies [37], is pursued through the incorporation of more formalization in CMs and through the analysis of the relations among concepts. Parts of the method presented here are based on a concept-sense disambiguation (CSD) algorithm defined by Simón *et al.* [33], and on WordNet [23]. The CSD algorithm tries to assign the most rational sense of a given concept in the CM, using WordNet, *contextual analysis* and *domains information*. The algorithm explores the context in which a given concept appear in the CM and try to determine a corresponding, similar context in WordNet, using the *synsets* (see below) of the concept at issue. A similar contextual analysis is carried out with the gloss.

In fact, in addition to the CMs to be formalized, two external knowledge sources will be used in this article: WordNet [23] and a CM repository. *WordNet* is a lexical knowledge base, whose basic structure is the synset. Synsets form a semantic network and are interconnected among themselves by several types of relations, some of which are used in the algorithm presented, such as *hypernymy-hyponymy* (class/subclass) and *meronymy-holonymy* (part/whole). The synsets define the meanings of a word, which, in the case of polysemy, can be found in various synsets. WordNet can be used as an ontology if its links are associated to a formal semantics. To work with CMs in languages other than English, EuroWord-Net is used. EuroWordNet, which was a European resources and development project supported by the Human Language Technology sector of the Telematics Applications Programme, is a multilingual database with wordnets for several European languages (Dutch, Italian, Spanish, German, French, Czech and Estonian). The *CM repository* used here is ServiMap [36], which stores several CMs of different domains (including the environmental domain). All CMs in ServiMap are constructed using the CM editor Macosoft [36].

1.4. Ontologies and OWL DL

In Artificial Intelligence, Ontologies were introduced to share and reuse knowledge. They provide the common reference frame for communication languages in distributed environments (such as multi-agent systems or the semantic Web) and a formal description for automatic knowledge processing. Several languages have been defined to implement them and OWL [37] is the latest, standardized ontology language. OWL is based on RDF and RDFS, and includes three specifications, with different expressiveness levels: OWL Lite, OWL DL and OWL Full. The code obtained by the method described in this article is generated according to OWL DL specifications. OWL DL is so named due to its correspondence with description logics. *Description logic* (DL) is the name for a family of knowledge representation formalisms that represent the knowledge of a domain by first defining the relevant concepts of the domain (its terminology), and then using these concepts to specify properties of objects and individuals occurring in the domain [3]. The terminology specifies the vocabulary of a domain, which consists of concepts and roles, where the concepts denote individuals while roles denote binary relationships between individuals.

2. A Multi-Agent System for Knowledge Management

In this section a *concept map knowledge management system* (CMKMS) and a *multi-agent system* (MAS) for knowledge management, which combines CMs, ontologies and WordNet, are presented. The general structure of the CMKMS is designed using the INGENIAS agent-oriented software-engineering methodology [28] [14], which is outlined in section §2.1. Then, the CMKMS is presented using agents as design metaphor.

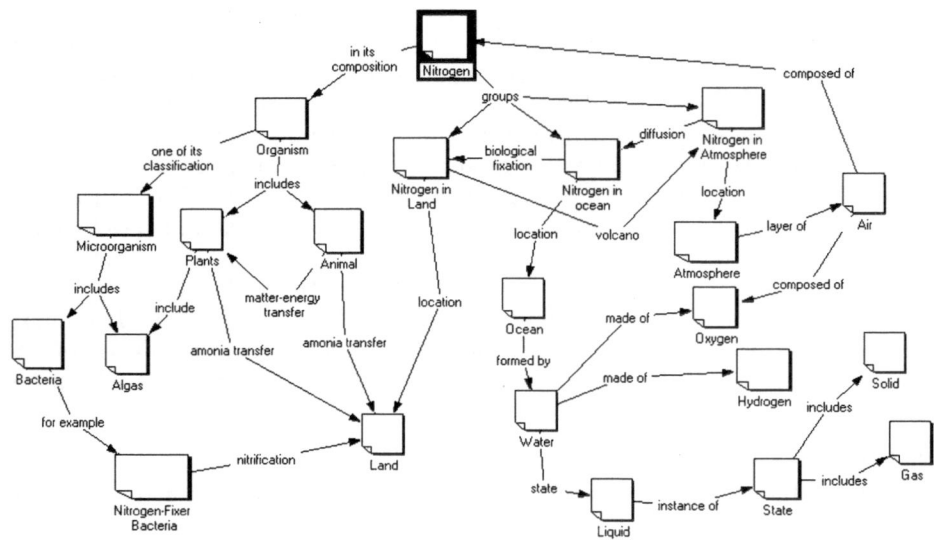

FIGURE 1. Concept map of the nitrogen cycle.

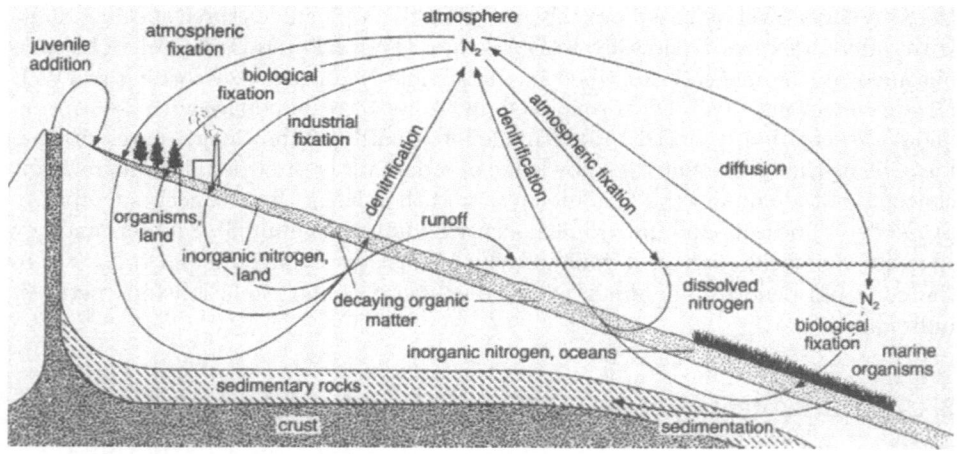

FIGURE 2. Representation of nitrogen cycle obtained from Jones et al. [16].

2.1. INGENIAS: An Agent-Oriented Software-Engineering Methodology

Although agent-oriented software engineering is still a young discipline, various methodologies have been proposed [14], among which the INGENIAS methodology is one of the most complete, covering the most important phases of the agent-oriented software development process. In order to develop the CMKMS, some of the advantages of INGENIAS (such as: software and methodological support, and support for conventional objects [27]) are very relevant, because CMKMS comes from GECOSOFT [36], an object-oriented software. Other aspects of INGENIAS (such as: support for open systems, dynamic structure, mobile agents and ontologies [27]) are less relevant to CMKMS. Also, *the best* methodology depends on the target application [27], and the aspects of the domain under consideration guided us to select INGENIAS as a good methodology in order to develop the CMKMS. It models many important aspects of the agent paradigm, such as intelligence, communication, environment of the system, goals, tasks and resources. Additionally, it allows the use of many object-oriented concepts [28].

A MAS is expressed in terms of particular concepts of the agent paradigm, such as roles, agents and groups. These concepts are now briefly explained, following the definition given by INGENIASs authors [28]:

- Agent: A program that follows the rationality principle and processes knowledge. The rationality principle states that an agent will execute actions in order to achieve its goals. The behavior of an agent is defined through three components: mental state (an aggregation of mental entities such as goals, believes, facts, and compromises), mental state manager (including operations to create, destroy, and modify mental entities), and mental state processor (describing the evolution of the mental state, in terms of rules and planning).
- Organization: It describes the framework where agents, resources, tasks, and goals coexist. It is defined by its structure, functionality, and social relationships. The structure defines a decomposition of the MAS in groups.
- Group: It may contain agents, roles, resources, or applications.
- Interaction: Exchange of information or requests between agents, or between agents and human users. It requires the definition of actors in the interaction (initiator and collaborators), interaction specification (how the interaction is constructed, mental attitudes and actions of agents), context of the interaction and nature of the interaction (attitude of interaction participants).

INGENIAS considers five different starting points for the development:

- Environment: Identification of other software which will coexist with the MAS.
- Interaction: Identification of interactions among agents, including direct associations with tasks and agent mental states.
- Organization: Top-down system-subsystem identification. Interfaces of systems and subsystem are identified as roles belonging to groups of agents. Interface operations are identified as tasks and workflows.

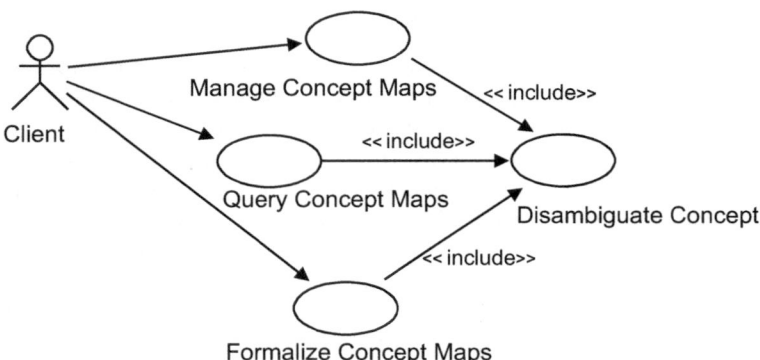

FIGURE 3. MAS-CMKMS use case diagram.

- Agents: Identification of system agents. This approach may produce a situation where everything is an agent. It is suggested to start with roles and, later on, assign them to agents.
- Tasks and goals: Identification of a set of tasks (and their expected output) and goals that the system has to execute or reach.

A detailed description of INGENIAS methodology can be found in Pavón *et al.* [28]. Other agent-oriented methodologies are presented by Henderson and Giorgini [14].

2.2. General Description of CMKMS

CMKMS is a MAS for knowledge management, which combines CMs [25], ontologies [37] and WordNet [23], and extends GECOSOFT [36]. The most important functionalities of CMKMS are the following ones:

- To manage concepts maps, images, presentations, documents, and other resources attached to concepts maps. This includes operations to download, upload, retrieve, delete CMs, and search the concept map repository.
- To allow complex operations using CMs in order to obtain new CMs. These operations allow the combination and selection of CMs, or parts of them.
- To obtain the most rational meaning of a concept in a concept map, solving the complex problem of dealing with ambiguous phrases or words.
- To obtain an ontological representation of concept maps, resulting in files in the OWL format, which may be used not only by human, but also by software agents.

INGENIAS employs *use case diagrams* to represent the external view of a system. Figure 3 presents this external view for CMKMS.

The diagram in Figure 3 is an external view and does not reveal the details of the metaphors and technologies used to implement the system. Next sections are focused on the internal view of the CMKMS as a MAS.

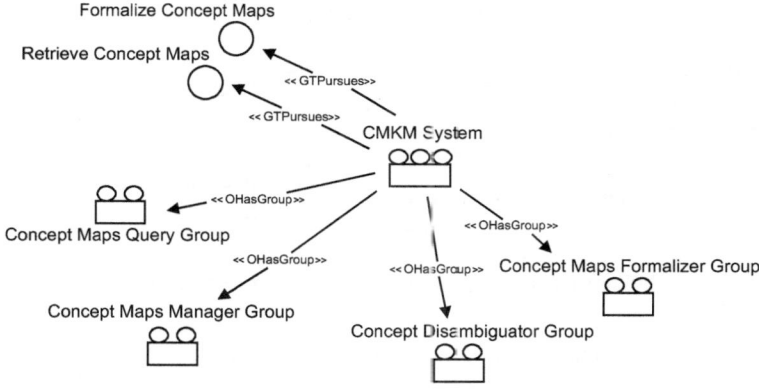

FIGURE 4. MAS-CMKMS system architecture.

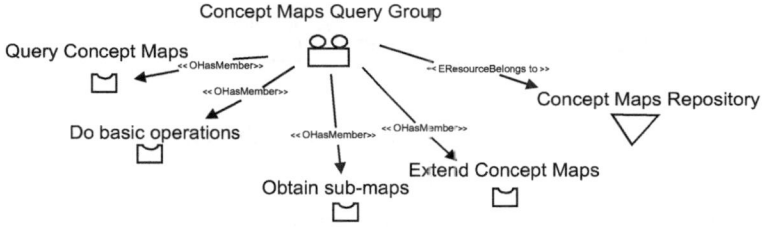

FIGURE 5. CM Query Group.

2.3. Internal Architecture

The underlying structure of the MAS, representing the internal structure of the MAS-CMKMS is presented in Figure 4, which represents CMKMS as a system with two main objectives: formalize and retrieve concepts maps, which summarize the functionality described in the previous section. The internal structure of CMKMS is composed of four groups: CM Manager Group, CM Query Group, Concept Disambiguator Group, and CM Formalizer Groups.

The objective of retrieving CMs is pursued by the CM Manager Group and the CM Query Group. The CM Manager Group is simpler than the other three groups. It deals with the traditional functions of CM repository management. These functions are similar to the ones integrated in GECOSOFT [36]. The other groups are described in more detail in Figure 5, 6 and 7.

Figure 5 describes the organization of the CM Query Group. It deals with four different roles: Query Concept Maps, Do basic operations (*e.g.*, union and intersection of concept maps), Obtain sub-maps, and Extend Concept Maps. The Query Concept Maps role acts as a coordinator for the other three roles. These

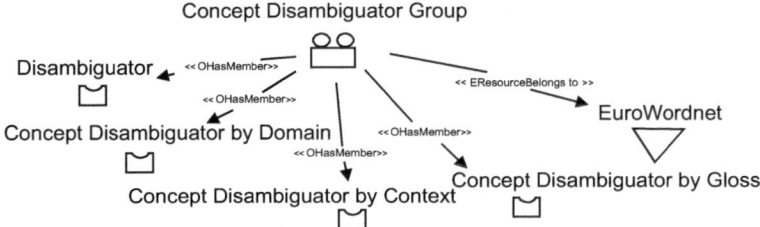

FIGURE 6. Concept Disambiguator Group.

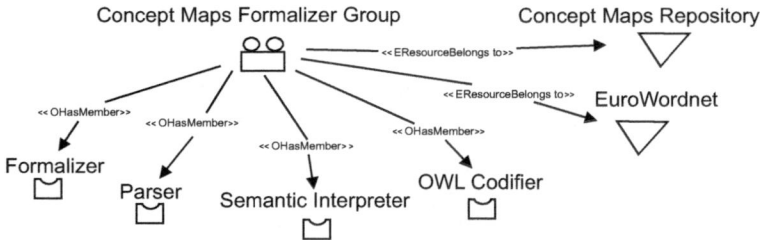

FIGURE 7. CM Formalizer Group.

three roles operate over the CM Repository, which is considered as a resource according to INGENIAS. Each one of these three roles is in charge of operations over concept maps in order to obtain new maps, as combination of some of them. For instance, one of the operations allows obtaining concept maps which extend a given map with other propositions about a particular concept. A detailed description of the operation allowed over the concepts map repository is presented in section §2.5.

Figure 6 shows the organization of the Concept Disambiguator Group. It contains four roles. The Disambiguator role coordinates the work of the other three roles, which correspond to different types of disambiguation: by domain, by context, and by gloss. Each one of these roles is responsible of obtaining the most rational meaning of each ambiguous concept in a CM. Each one uses different heuristic criteria in order to make a decision. Each one of these heuristics is explained in section §2.6. The Disambiguator uses two recourses: the CM Repository, and the lexical database EuroWordNet.

Figure 7 shows the organization of the CM Formalizer Group. It contains four roles. The Formalizer role coordinates the work of the other three roles, which deal with specific steps in the process toward obtaining a formal OWL codification of a concept map. An algorithmic description of all the processes in presented in section §2.7, as a description of the responsibilities of all the roles involved in this group.

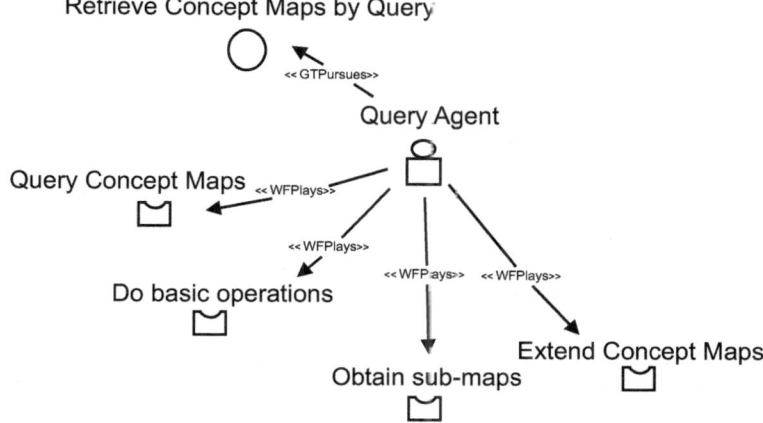

FIGURE 8. Query agent description.

The other roles in this group are: Parse, Semantic Interpreter, and OWL Codifier. In order to accomplish their respective task, this group uses two resources: the CM Repository, and the lexical database EuroWordNet. The Formalizer is also responsible of the interaction with the Disambiguator role of the Concept Disambiguator Group. This interaction does not appear in Figure 7. It is explained in section §2.4 along with other aspects of each agent.

2.4. General Description of the Agents

Each of the three groups described in the previous section is implemented as a single agent, playing different roles. This internal implementation is presented in Figures 8, 9 and 10.

Figure 8 explains that the Query agent is responsible of all the roles described for the CM Query Group. This agent implements complex operations, which are based on graph theory, to obtain new concepts maps. These operations combine and filter the concepts maps stored in the CM repository. A detailed description of the operations implemented is presented in section §2.5.

Figure 9 presents the description of Concept Disambiguator Agent, with the implementation of all the roles included in the Concept Disambiguator Group. It also expresses that the Concept Disambiguator Agent has a Mental Processor and a Mental Manager for performing the complex tasks associated to these roles. Associated states include those of the disambiguation process, according to the state in which a concept is at each moment, such as ambiguous, disambiguated by domain, disambiguated by context, or disambiguated by gloss. A detailed description of this complex processing is presented in algorithmic style in section §2.6. Each role is part of the algorithm of concept disambiguation.

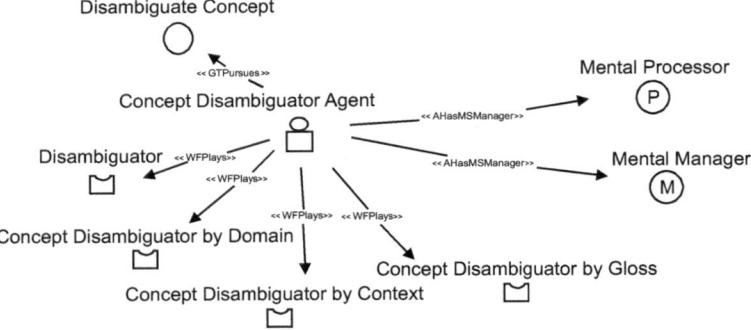

FIGURE 9. Concept Disambiguator Agent description.

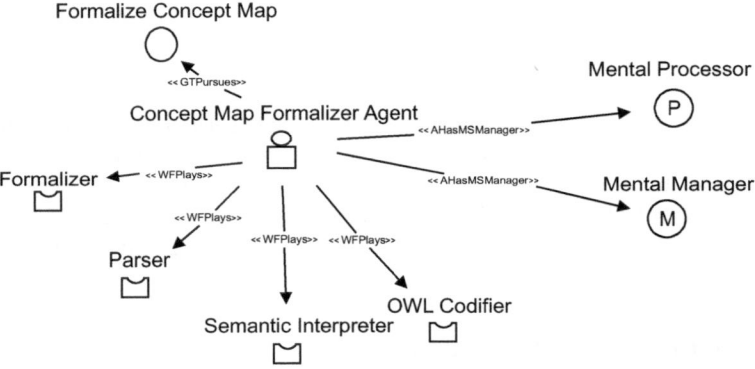

FIGURE 10. CM Formalizer Agent description.

Figure 10 presents the description of the CM Formalizer Agent and the roles it implements (all the ones included in the CM Formalizer Group). It also expresses that the CM Formalizer has a Mental Processor and a Mental Manager for performing the complex tasks associated to these roles. The states here evolve, starting from single propositions identified in the CMs towards specific OWL code that formalizes these propositions. A detailed description of this complex processing is presented in a descriptive style in section §2.7. Each role is expressed as a group of production rules that are processed in a chaining mechanism in order to carry out the whole process of concept map formalization.

As already mentioned, the Formalizer role is not only to coordinate the work of the other roles in the group, but also to interact with the Disambiguator. Being the Disambiguator role implemented by the Disambiguator Agent, it is necessary to describe this interaction, and this is done in Figures 11 and 12.

FIGURE 11. Description of the interaction associated to the *Formalize CMs* use case.

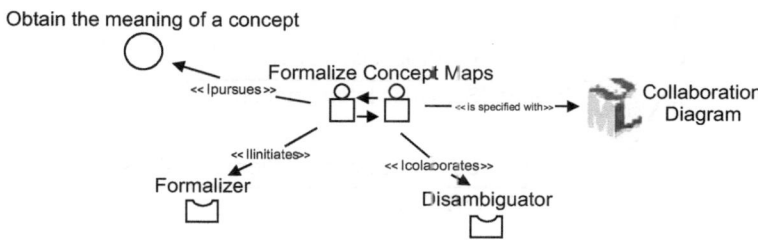

FIGURE 12. Detailed description of the interaction associated to the *Formalize CMs* use case.

Figure 11 describes that, as part of the task to *Formalize CMs*, it is necessary an interaction to *Obtain the meaning of a concept*. This interaction is also related to the *include* relation between *Formalize CMs* and *Disambiguate Concept* (not shown). Figure 12 describes that the interaction is initiated by the Formalizer with the objective of obtaining the meaning of a concept. The Disambiguator (which is implemented by another agent) collaborates with the Formalizer in order to accomplish this objective.

The same interaction is described using a UML collaboration diagram, which is presented in Figure 13.

Figure 13 presents the collaboration diagram that describes the sequence of the interactions between the roles involved in the *Formalize Concepts Maps* use case, including the interaction with the Disambiguator. This interaction is repeated for each concept. For each ambiguous concept, the Formalizer sends to the Disambiguator a request for disambiguation. The Disambiguator responds with the (hopefully) most rational meaning of the concept. A detailed description of the whole process is presented in section 2.7.

2.5. Query Agent

The Query agent implements the most complex part of the retrieval capabilities supported by CMKMS. This section gives more details about the operations implemented by this agent.

2.5.1. Retrieval Process: Concept Maps Query Algebra.

In this section, the operations for obtaining knowledge from a CM repository are presented. The whole set of operations are presented as part of a *concept maps query algebra* (CMQL). CMQL is a mathematical description of the operations available using *graph theory*

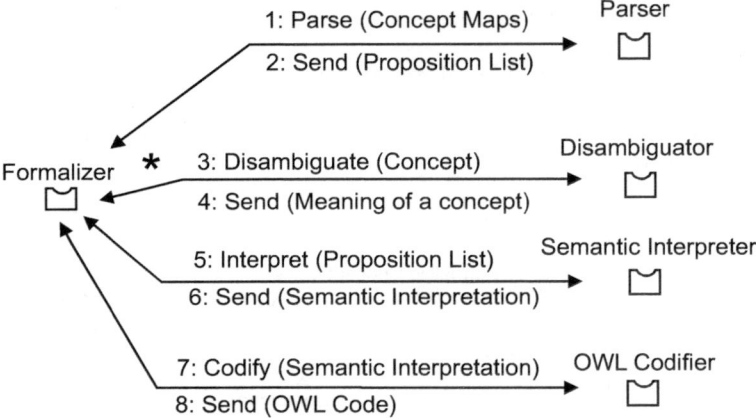

FIGURE 13. Collaboration diagram describing the interaction associated to the *Formalize CMs* use case.

and *set theory*. The result of the application of each operation is the automatic construction of a new CM. The Query agent retrieves information (*concepts and propositions*) from a repository of CMs, through the following operations:

1. union of a CM set;
2. intersection of a CM set;
3. closed sub-map, guided by a concept set;
4. open sub-map, guided by a concept set;
5. open sub-map of radio R, guided by a concept set;
6. closed extension of a CM, guided by another CM and a concept set;
7. open extension of a CM, guided by another CM and a concept set;
8. open extension of radio R of a CM, guided by a concept set.

For the information retrieval in the repository, the agent uses one or several of the query operations included in CMQL. The information that is retrieved from the repository is formed by *concepts* and *propositions*, and is expressed as the automatic construction of a new CM, in which those concepts and propositions are related. The query operations defined in CMQL are formalized in terms of the combination of graph theory and set theory, and may have as input one or more CMs (and a concept set in some cases) (as shown in Table1). The CM is represented as a *directed graph* [15] , that is, $G = (V, E)$, where V is the set of vertices (*concepts*) and E the set of directed edges (*propositions*). This allows taking advantage of the operations that have been defined in both areas (graph theory and set theory) for the automatic processing of CMs.

2.5.2. Concept Maps Query Algebra: Definitions and Operations. The CM query operations allow automatically obtaining a new CM, which can be edited later by the user, from knowledge represented in other CMs. This is a novel contribution

Basic definitions:
M^x is a concept map, $M^x = (C^x, P^x)$; c is a concept;
C^x and CS are concept sets;
P^x is a proposition set, $P^x = \{\ldots,(c_o, l\text{-}p_j, c_d), \ldots\}|$ $l\text{-}p$ is a *linking-phrase* and $c_o, c_d \in C^x$;
CMS is a set of concept maps.

Query operations	Expression	Results
Union of a CM set	$UM(CMS) = \cup\, M^i\ (M^i \in CMS)$	A new CM formed by: all *concepts* and *propositions* represented in the CMs included in CMS.
Intersection of a CM set	$IM(CMS) = \cap\, M^i\ (M^i \in CMS)$	A new CM formed by: the *concepts* presented in all CMs in CMS, and the *propositions* in which they are related.
Closed sub-map, guided by a concept set	$SM^-(M^x, CS) = M^x \cap M^y$ $(M^y = (CS,\{\}))$	A new CM formed by: • the common *concepts* between CS and M^x; and • the *propositions* in M^x in which they are related.
Open sub-map, guided by a concept set	$SM^+(M^x,CS) = M^1 = (C^1,P^1)$ $(P^1 =\{(c_o,l\text{-}p_j,c_d)\ \|$ $(c_o,l\text{-}p_j,c_d)\in P^x, c_o \in (CS\cup C^x) \vee$ $c_d \in (CS\cup C^x)\}$, $C^1 = \{c_i(c_i,l\text{-}p_j,c_d) \in P^1\} \cup$ $\{c_i(c_o,l\text{-}p_j,c_i) \in P^1\}\ )$	A new CM formed by: • the *concepts* in CS and their neighbors in M^x (two concepts are neighbors if they are related by a proposition); • the *propositions* in M^x in which the previous concepts are related.
Open sub-map of radio R, guided by a concept set	$SM^{+,R}(M^x, CC) =$ $\begin{cases} SM^+(M^x, CC) & \text{if } R = 1 \\ SM^{+,R-1}(M^x, C1), \\ M1 = (C1, P1) \\ = SM^+(M^x, CC) & \text{if } R > 1 \end{cases}$	A new CM formed by: • the common *concepts* between CS and M^x and all *concepts* in M^x to which a path with length $\leq$ R can be created from some concept in CS; • the *propositions* in M^x in which those concepts are related.
Closed extension of a CM, guided by another CM and a concept set	$Ext^-(M^x, M^y, CS)$	A new CM formed by: • the *concepts* in M^x and the *concepts* included in the CM obtained from $SM^-(M^y,CS)$; • the *propositions* in M^x and the *propositions* included in the CM obtained from $SM^-(M^y, CS)$.
Open extension of a CM, guided by another CM and a concept set	$Ext^+(M^x, M^y, CS)$	A new CM formed by: • the *concepts* in M^x and the *concepts* included in the CM obtained from $SM^+(M^y,CS)$; • the *propositions* in M^x and the *propositions* included in the CM obtained from $SM^+(M^y, CS)$.
Open extension of radio R of a CM, guided by another CM and a concept set	$Ext^{+,R}(M^x, M^y, CS)$	A new CM formed by: • the *concepts* in M^x and the *concepts* included in the CM obtained from $SM^{+,R}(M^y,CS)$; • the *propositions* in M^x and the *propositions* included in the CM obtained from $SM^{+,R}(M^y,CS)$.

TABLE 1. CM query operations included in CMQ.

with respect to current retrieval proposals, in which *concepts* and *propositions* are retrieved independently and have to be integrated by the user [8]. With the proposed method, CMs developed by persons focused on different aspects of a domain can be integrated, as in the case of *master maps* [24], which can be automatically obtained using the operation *Union of a CM set*.

2.6. Concept Disambiguator Agent

The disambiguation process on a CM, accomplished by the Concept Disambiguator Agent (see Figure 9), is now presented in algorithmic style. It is carried out through a *concept sense disambiguation* algorithm (CSD), which is formally described is this section. The CSD algorithm comprises five steps: *preparing the CM, selecting a set of CM domains (D_{cm}), disambiguating by domain, disambiguating by context* and *disambiguating by gloss*. These steps are executed sequentially on a CM and the order was defined experimentally to obtain a more efficient processing. The *disambiguation by domain*, for instance, required fewer queries to WordNet than the *disambiguation by context* and the precision obtained in the process is better; the *gloss* is included in the CSD algorithm as an alternative if some concepts cannot be disambiguated by domain or context. In the process, concepts, when disambiguated, are added to a set of non-ambiguous concepts, together with their sense.

For this study, CM in both English and Spanish are used. Along the rest of the article, to represent the English translation of the Spanish terms used, the following notation will be employed: *español* ("Spanish").

Before describing the *steps* of the algorithm, let us consider the following basic data:

- C is the set of concepts (c) in the CM;
- $S(c)$ is the set of *synset* (s) corresponding to concept c; e.g., the synset{*ser_vivo#1, ser#1, organismo#1*} corresponding to concept *Organismos* ("organism");
- $S(C)$ is the set of *synsets* corresponding to all concepts in C;
- $D(s)$ is the set of domains (d) associated to s; e.g., the domains {*Chemistry, Physics*} associated to the synset{*nitrógeno#1, número_atómico_7#1*};
- $D(c)$ is the set of domains associated to the set of *synsets of c*;
- $D(C)$ is the set of domains associated to the set of *synsets* of all concepts in C;
- $CSD(C, d)$ is the subset of concepts in C which have at least one *synset* associated to the domain d: $CSD(C, d) = \{c_i \mid c_i \in C, d \in D(c_i)\}$; e.g., $CSD(\{Nitrógeno, Atmósfera, Tierra\}, Physics) = \{Nitrógeno, Atmósfera\}$;
- $OF(d,C)$ is the *occurrence frequency* of domain d in the *synsets* of the concepts in C:

$$OF(d, C) = \frac{|CDS(C,d)|}{|C|}$$

- $D_{ch}(D)$ is the set of child domains of the domains included in D according to the taxonomy of WordNet; e.g., $D_{ch}(\{Biology,\ Geography\})=\{Biochemistry,\ Anatomy,\ Physiology,\ Genetics,\ Topography\}$;
- $D_p(D)$ is the set of parent domains of the domains included in D according to the taxonomy of WordNet; e.g., $D_p(\{Biology,\ Geography\})=\{Pure\ Science,\ Earth\}$;
- T is a coefficient that defines the percentage of concepts in the CM, to be considered for determining the CM's domains;
- $Context_{cm}(c,\ r)$ is the set of neighbor concepts of a given concept c within a radius r (measured as arcs between two concepts) in the CM and the words (*nouns*, *adjectives* and *verbs*) extracted from the *linking-phrases* used in the proposition in which these concepts are related;
- $Context_{wn}$ $(s,\ L,\ C)$ is the set formed by paths between *synset* s and other synsets s' in WordNet, with a maximum length of L (measured as arcs between two synsets) from s, such that $s' \in\ S(C)$ and using *hyperonymy*, *meronymy* and *gloss* relations; e.g. $Context_{wn}$ $(\{agua\#,\ H2O\#1\},\ 2,\ \{Hidrógeno,\ Oxígeno\})=$ $\{(\{hidrógeno\#1,\ número\_atómico\_1\#1\},\ 1,\ 1),\ (\{número\_atómico\_8\#1,\ O\#1,\ oxígeno\#1\},\ 1,\ 1),...\}$, from the paths: $\{agua\#,\ H2O\#1\}$ *has_mero_madeof* $\{hidrógeno\#1,\ número\_atómico\_1\#1\}$ and $\{agua\#,\ H2O\#1\}$ *has_mero_madeof* $\{número\_atómico\_8\#1,\ O\#1,\ oxígeno\#1\}$;
- $w(Context_{wn}(s,\ R,\ C))$ represents the weight of a sense s to disambiguate a concept c:

$$w(Context_{wn}(s, L, C)) = \sum_{k=1}^{|Context_{wn}(s,L,C)|} \frac{\alpha_k}{l_k}$$

- where l_k is the length of the path (k) and α_k is the number of concepts in C with some *synset* in k;
- $gloss(s)$ is the set of words included in the gloss of the synset s in WordNet.

Step 1. Preparing the CM

Extract all concepts (c_i) and the propositions they belong to from the CM; the *proposition* set (PS) and *concept* set (CS) are created. From CS, the following sets are created[1]:

- the *non-ambiguous concept* set NACS $= \{(c_i \mid c_i \in CS, \mid S(c_i)\mid = 1, s_{ij})\}$;
- the *unknown concept* set UCS $= \{c_i \mid c_i \in CS, |S(c_i)| = 0\}$;
- the *ambiguous concept* set ACS $= \{c_i \mid c_i \in CS, |S(c_i)| > 1\}$.

---

[1]The senses of the concepts are found using WordNet, after applying a morphological transformation where needed; using the FreeLing tool [1] . The transformation simply consists in obtaining the singular form of the concept if it appears in plural.

Step 2. Selecting a set of CM domains(D_{cm})
    r = 1;
    While ($|\text{Context}_{cm}(root\ concept,\ r)| <$ T * |CS|){r = r+1};
    DS = D($\text{Context}_{cm}(root\ concept,\ r)$);
    DS_{max} = {d_{max} | $d_{max} \in$ DS, $\forall d_i \in$ DS OF(d_{max}, $\text{Context}_{cm}(root\ concept,$
    r)) $\geq$ OF(d_i, $\text{Context}_{cm}(root\ concept,\ r)$))};
    D_{cm} = $DS_{max} \cup$D($root\ concept$).

Step 3. Disambiguating by domain
    For each $c_i \in$ ACS
    c_i is considered disambiguated by s_{ij} if one of the follow conditions is true:
1. $|\{s_{ij}|s_{ij} \in$ S(c_i), $|$D(s_{ij})$\cap$$D_{cm}|> 0\}| = 1$;
2. $|\{s_{ij}|s_{ij} \in$ S(c_i), $|$D(s_{ij})$\cap$$D_{ch}(D_{cm})|> 0\}| = 1$ and
   $|\{s_{ij}|s_{ij} \in$ S(c_i), $|$D(s_{ij})$\cap$$D_{cm}|> 0\}| = 0$;
3. $|\{s_{ij}|s_{ij} \in$ S(c_i), $|$D(s_{ij})$\cap$$D_{p}(D_{cm})|> 0\}| = 1$ and
   $|\{s_{ij}|s_{ij} \in$ S(c_i), $|$D(s_{ij})$\cap$$D_{cm}|> 0\}| = 0$.
    If c_i is disambiguated, then NACS = NACS $\cup$ {(c_i, s_{ij})}, ACS = ACS – {c_i}.

Step 4. Disambiguating by context
    For each $c_i \in$ ACS
    r = 1; W_d = 0; S_d = {};
    repeat
    C_t = Context_{cm}(c_i, r);
    For each $s_{ij} \in$ S(c_i)
    if (w(Context_{wn}(s_{ij}, L, C_t)) > W_d), then
    S_d= {s_{ij}}; W_d = w(Context_{wn}(s_{ij}, L, C_t));
    else
    if (w(Context_{wn}(s_{ij}, L, C_t)) = w_d), then S_d = $S_d \cup${s_{ij}};
    r = r + 1;
    until ($|S_d| = 1 \lor |C_t| = $|CS|)
    If $|S_d| = 1$, then
    c_i is disambiguated with s_{ij}; NACS = NACS $\cup$ {(c_i, s_{ij})}, ACS = ACS – {c_i}.

Step 5. Disambiguating by gloss
    For each $c_i \in$ ACS
    r = 1; G_d = {}; S_d = {};
    repeat
    C_t = Context_{cm}(c_i, r);
    For each $s_{ij} \in$ S(c_i)
    if ($|\text{gloss}(s_{ij}) \cap C_t| > |G_d|$), then
    S_d= {s_{ij}}; G_d = gloss(s_{ij}) $\cap C_t$;
    else
    if ($|\text{gloss}(s_{ij}) \cap C_t | = |G_d|$), then S_d = $S_d \cup${s_{ij}};

r = r + 1;
until $(|S_d| = 1 \vee |\text{Context}_{cm}(c_i, r)| = |CS|)$
if $|S_d| = 1$, then
c_i is disambiguated with s_{ij}; NACS = NACS $\cup \{(c_i, s_{ij})\}$, ACS = ACS $- \{c_i\}$.

In this algorithm, steps 3, 4 and 5 are associated with the roles described as Concept Disambiguator by Domain, by Context and by Gloss presented in Figures 6 and 9. The interaction with EuroWordNet is an essential part of the disambiguation process and is included in the algorithm. The Disambiguator role is mainly concerned with the coordination of the work of the algorithm as a whole.

2.7. Formalizer Agent

In the formalization process, accomplished by the CM Formalizer agent (see Figure 10), each role is related to a group of production rules, that are processed in order to formalize a concept map, that is, to translate a CM into OWL format.

2.7.1. Obtaining OWL DL Ontologies from Concept Maps.
Important similarities between CMs and the ontologies coded in resource description framework (RDF)-based languages (such as OWL) can be identified, given that RDF is a metadata model formalized through triples (subject, predicate, and object) and CMs use the proposition structure (concept, linking-phrase, concept). The knowledge in OWL ontologies is expressed as *classes, properties* and *instances* [37] , while in CMs much of this formal and explicit specification does not exist, and has to be inferred. Nonetheless, some initial structural mapping between CMs and OWL can be easily established [33] :

- concepts correspond to: classes and instances;
- linking-phrases correspond to: properties, considering this as a binary relation between instances of classes in OWL [37].);
- propositions correspond to: classes and properties' restrictions and other OWL DL constructs.

The mapping and semantic inference leading to OWL DL coding, in this article, is carried out combining the analysis of:

- the syntax: the semantic inferred from the linking-words used in the proposition;
- the natural language: the semantic relations between the senses of two concepts included in the proposition in WordNet [23] ;
- the topology: section topology in which the proposition appears in the CM;
- the experience: the occurrence of similar relations between two concepts in the proposition in a external CM repository.

Some type of semantic relation, such as: relations between *class* and *subclass*, relations between *class* and *property*, relations between *class, property* and its *value*, relations between *class* and *instance*, can be inferred from certain *linking-words* used in CMs, in accordance with other experiences [6] [11] . A set of frequently used *linking-words* are defined and organized in four categories, according

to the semantics that can be associated to them and their correspondence with
the semantic relations in WordNet, such as: Classification (CC), Instance (IC),
Property (PC) and Property-Value (PVC) Categories [34].

In the mapping method, the CM under consideration is analyzed as a structured text; considering propositions as natural language sentences. A concept
sense-disambiguation algorithm [35] is used to infer the most rational sense for
all concepts in the CM. Once inferred a *synset* for each concept in a *proposition*,
the semantics of the CM relation among them can be inferred from the relation in
WordNet (if one exists).

2.7.2. Map-to-OWL Method.

2.7.2. Map-to-OWL Method. The Map-to-OWL method for obtaining the preliminary OWL DL ontology from a CM is organized in three phases: *preprocess*,
mapping and *codification*. The Formalizer agent implements the Formalizer role
and the other three roles represented in Figure 10: Parser, Semantic Interpreter,
and OWL Codifier. As explained in Figures 11, 12 and 13, the Formalizer agent is
also the initiator of the interaction with the Disambiguator agent, as it is needed
in the formalization process. In this article, the rules for CM-OWL mapping have
been improved with respect to the ones reported by Simón *et al.* [33].

Preprocess phase. The *parser* analyzes the CM, extracting propositions and their
parts (*concepts* and *linking-phrases*), and creates a proposition set (PS) and a
concepts set (CS) which includes all concepts in PS. The set PS have *(C_o, linking-
phrase, C_d)* as basic structure, where C_o is the origin concept and C_d is the destina-
tion concept. Besides, the *parser* extracts from the CM repository all propositions
in which at least one concept included in CS appear and creates, with these, a
second proposition set (PP-BC).

The *disambiguator* infers the most rational *sense* (in terms of WordNet's
synsets) of the concepts in the CM, using the algorithm reported by Simón *et al.*
[35], and analyzes the senses inferred according to the *hypernymy/hyponymy* and
meronymy/holonymy relations between *synsets* found in WordNet:

- A **PS_WN**$_{hype-hypo}$ set is created with:
  - the pair (C, C'), if the *synsets* of two concepts C and C' are directly related by a
    hypernymy; and
  - the pair (C', C), if the *synsets* of two concepts C and C' are directly related by a
    hyponymy.

- A **PS_WN**$_{mero-holo}$ set is created with:
  - the pair (C, C') if the *synsets* of two concepts C and C' are directly related by a
    has_meronym; and
  - the pair (C', C) if the *synsets* of two concepts C and C' are directly related by a
    has_holonymy.

- A **PS_WN**$_{mero-holo-type}$ set is created with:
  - the triple (C, C', type of relation) if the *synsets* of two concepts C and C' are directly related by some type of *meronymy* WordNet's relation (e.g. *has_mero_madeOf*), different from *has_meronym*; and
  - the triple (C', C, type of relation) if the *synsets* of two concepts C and C' are directly related by some type of *holonymy* WordNet's relation (e.g. *has_holo_madeOf*), different to *has_holonym*.

Mapping phase. A set of heuristic rules (if-then) are defined for mapping between the propositions (P) included in PS to OWL DL constructs. These rules uses the set of *linking-phrases* (*lp*) included in the four categories mentioned earlier and the sets generated in the *preprocess phase*. These rules are described using the set theory. The *semantic interpreter* applies the set of heuristic rules defined to the propositions obtained by the *parser* until all rules are applied. A salience value is attributed to each rule to guide their execution order.

Rules for simple class specification. Rules R-1 and R-2 are defined to infer hierarchical relations and simple classes in the CM using the *lp* included in CC and the *hypernymy/hyponymy* WordNet's relations (in the case of those *lp* cannot be used); where the *lp* used in the proposition is learned (see R-2). These hierarchical relations can be established through two types of *lp*: to denote a descending hierarchical relation, such as *includes, type of* (included in CC) or to denote ascending hierarchical relations, such as *is a, a type of* (included in CC^{-1}). A S_CS set of the pairs (C, C'), where C is the class and C' is the subclass, is generated.

R-1 (salience = 8):
  (a) If $P = (C_o, lp, C_d) \in PS \land lp \in CC$, then $PS = PS - \{P\}$, $S\_CS = S\_CS \cup \{(C_o, C_d)\}$.
  (b) If $P = (C_o, lp, C_d) \in PS \land lp \in CC^{-1}$, then $PS = PS - \{P\}$, $S\_CS = S\_CS \cup \{(C_d, C_o)\}$.

R-2 (salience = 8):
  If $P = (C_o, lp, C_d) \in PS \land lp \notin CC \land lp \notin IC \land (C_o, C_d) \in PS\_WN_{hype-hypo}$, then $PS = PS - \{P\}$, $S\_CS = S\_CS \cup \{(C_o, C_d)\}$, $CC = CC \cup \{lp\}$.

Rules for complex class specification. Rules R-3 to R-5 are defined to infer *complex classes* in the hierarchical representation of the CM, corresponding to *union class* and *intersection class* in OWL. The following sets are generated:

- *Intersection$_{classes}$* is a set of pairs *(CI, {... C_i...})*, where *CI* is the *intersection class* of the classes included in *{... C_i...}*.
- *Intersection$_{class-property}$* is a set of triples *(CI, C, Pr, V)*, where *CI* is the *intersection class* between the class *C* and the previously inferred property *Pr* and *V* is its value.
- *Union* is a set of pairs *(CU, {... C_i...})*, where *CU* is the union class of the classes included in *{... C_i...}*.

The rules are carried out through the analysis of the CM's topology and the use of some previously generated sets.

R-3 (salience = 6):
   If $U = (CU, \{C_i \mid (CU, C_i) \in S\_CS\}) \wedge U \notin$ *Union*, then *Union* = *Union* $\cup \{U\}$.

R-4 (salience = 6):
   If $I_C = (CI, \{C_i \mid (C_i, CI) \in S\_CS\}) \wedge I_C \notin$ *Intersection*$_{classes}$, then *Intersection*$_{classes}$ = *Intersection*$_{classes}$ $\cup \{ I_C \}$.

R-5 (salience = 2):
   If $I_{C-Pr} = (CI,\ C \mid ((C,\ CI) \in C\_CS,\ (C',\ CI) \notin C\_CS),\ Pr \mid (CI,\ Pr,\ V) \in C\_CPV_{hasValue},\ V) \wedge I_{C-Pr} \notin$ *Intersection*$_{class-property}$, then *Intersection*$_{class-property}$ = *Intersection*$_{class-property}$ $\cup \{I_{C-Pr}\}$.

Rule for instance specification. Rule R-6 is defined to infer relations between classes and instances in the propositions of the CM, using the *lp* included in IC. These relations can be established through two types of *lp*: to denote a descending relation, such as *has example, has instance* (included in *IC*); or to denote ascending relations, such as *instance of* (included in IC^{-1}). An *S_CI* set of the pairs *(C, I)*, where C is the class and I is an instance of C, is generated.

R-6 (salience = 8):
   (a) If $P = (C_o,\ lp,\ C_d) \in PS \wedge lp \in IC$, then $PS = PS - \{P\}$, $S\_IC = S\_IC \cup \{(C_o, C_d)\}$.
   (b) If $P = (C_o,\ lp,\ C_d) \in PS \wedge lp \in IC^{-1}$, then $PS = PS - \{P\}$, $S\_IC = S\_IC \cup \{(C_o, C_d)\}$.

Rules for property specification. Rules R-7 and R-8 are defined to identify two types of property relations in the CM: the relation between class and property and the relation between class, property and its value. These rules are carried out through the analysis of the *lp* included in PC or PVC, and the *meronymy/holonymy* WordNet's relations (in case those *lp* cannot be used; the *lp* used in the proposition is then learned). The following sets are generated: *S_CP* of the pairs *(C, Pr)*, where C is a class and Pr is a property of C; and *S_CPV* of the triples *(C, Pr, C')*, where C is a class, Pr is a property of C and C' is a range of values refers to the values of *Pr*.

R-7 (salience = 8):
   If $P = (C_o,\ lp,\ C_d) \in PS \wedge (lp \in PC \vee (C_o, C_d,) \in PS\_WN_{mero-holo})$, then $PS = PS - \{P\}$, $S\_CP = S\_CP \cup \{(C_o, C_d)\}$, $PC = PC \cup \{lp\}$.

R-8 (salience = 8):
   (a) If $P = (C_o,\ lp,\ C_d) \in PS \wedge lp \in PVC$, then $PS = PS - \{P\}$, $S\_CPV = S\_CPV \cup \{(C_o, lp, C_d)\}$;
   (b) If $P = (C_o,\ lp,\ C_d) \in PS \wedge (C_o, C_d, type) \in PS\_WN_{mero-holo-type}$, then $PS = PS - \{P\}$, $S\_CPV = S\_CPV \cup \{(C_o, property^2, C_d)\}$, $PVC = PVC \cup \{lp\}$.

Rule for property restriction specification. Rule R-9 is defined to infer property restrictions in the CM corresponding to *hasValue* in OWL, and a $S\_CPV_{hasValue}$ set of the triples *(C, Pr, C')* is generated, where C is a class, Pr its property and C' the range of values of the property *Pr*.

R-9 (salience = 6)::
   If $(C, Pr, I) \in S\_CPV \wedge \exists(c, I) \in S\_CI$, then $S\_CPV_{hasValue} = S\_CPV_{hasValue} \cup \{(C, Pr, I)\}$.

---

[2]This is the property corresponding to type in PVC, for example, *hecho de* ("made of") in case of *has_mero_madeof* relation.

Rules for property characteristics specification. Rules R-10 and R-11 are defined to infer *property characteristics* in the CM, corresponding to *functional property* and *symmetric property* in OWL, using the CM repository, some previously generated sets, and *synonymy*, *meronymy* and *hypernymy* WordNet's relations. Sets $Pr_{symmetric}$ of symmetric properties and $Pr_{functional}$ of functional properties are generated.

R-10 (salience = 2):
> If $(C, Pr, I) \in S\_CPV \wedge \forall C ((C, lp, C') \in PP\text{-}BC \mid (lp = Pr$ or they are synonyms in WordNet) $\wedge$ (lp = Pr or they are synonyms in WordNet)), then $Pr_{functional} = Pr_{functional} \cup \{Pr\}$.

R-11 (salience = 6)::
> If $(C, Pr, V) \in S\_CPV \wedge (\forall V((V, lp, C') \in PP\text{-}BC \mid (lp = Pr$ or they are synonyms in WordNet) $\wedge$ (C' = C or they are synonyms in WordNet))) $\vee$ (Pr is meronym of c and hyperonym of C in WordNet), then $Pr_{symmetric} = Pr_{symmetric} \cup \{Pr\}$.

Codification phase. The *OWL codifier* uses the sets generated by the *semantic interpreter* and writes out the corresponding OWL constructs according to W3C recommendation [37], considering the mapping conventions shown in Tables 2–5 (see pp. 166–167).

3. Application to Environmental Domains

Environmental education is a process in which individuals get conscious about their environment and acquire knowledge and values so that they will be able to act upon and solve environmental problems. CMs have been widely used for knowledge modeling in environmental domains [13] [19] [21] [32], for improving the quality of teaching, learning and assessing environmental knowledge [12] [17] [18] [31] [38] [39] [40], for improving the design of environmental monitoring processes [29], for obtaining a structured organization of environmental aspects [22], for creating global learning spaces for sustainable development [30], and for increasing text comprehension related to environmental education [10]. In these experiences, several environmental-knowledge management activities using CMs are carried out, such as representation, acquisition, integration and organization of knowledge.

CMKMS allows increasing the capabilities for computational processing of CMs in a knowledge management system, according to knowledge retrieval from a CM repository, sense-disambiguation of the concepts represented in CMs, and OWL formalization of the semantics of the CMs. The latter aspect facilitates the use of generated and managed knowledge to not only people but also other software agents.

3.1. Knowledge Retrieval by the Query Agent

As an example of knowledge retrieval, three query operations (*intersection*, *union* and *open sub-map*) are performed by the Query agent on the sample CMs about environmental knowledge shown in Figure 14, considered as a CM repository. The CMs, result of the queries, are shown in Figure 15.

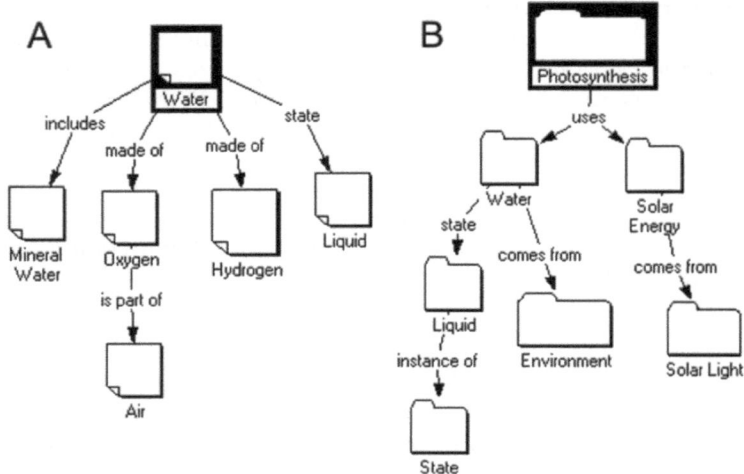

FIGURE 14. Sample repository of concept maps.

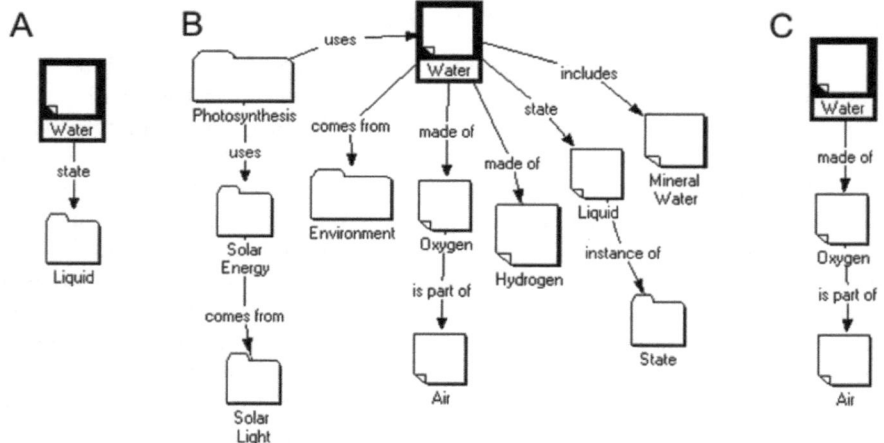

FIGURE 15. A: intersection of knowledge included in the CM repository, IM({Water, Photosynthesis}). B: union of knowledge, UM({Water, Photosynthesis}). C: open sub-map of the CM repository, guided by *Oxygen* and *Air* concepts: SM$^+$(UM({Water, Photosynthesis}),{Oxygen, Air}).

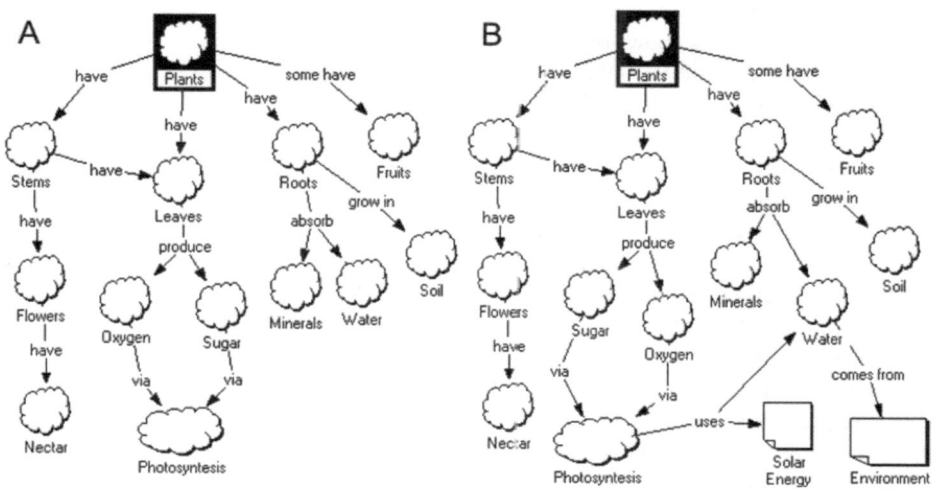

FIGURE 16. A: concept map about *Plants*. B: open extension with R=1 of *Plants* CM using the CM repository of Figure 14, guided by *Photosynthesis* and *Environment* concepts, $Ext^{+,1}$(Plants, UM({Water, Photosynthesis}), {Photosynthesis, Environment}).

As another example, a query operation (*open extension* with R=1) is performed by the Query agent to extend the CM shown in Figure 16.

The Query agent can be useful to obtain a conceptualization from a CM repository, which may be later translated into OWL by the Formalizer agent. In this way it is possible to formalize the informal knowledge of a CM into an ontology.

3.2. OWL Formalization (of a Conceptualization) by the Formalizer Agent

As an example of ontology formalization, we use the sample CM shown in Figure 15B, which was obtained from a *union* operation. This example represents a case in which a user needs an ontology about some environmental concepts; he does not have it available; but he knows about some knowledge related with the subject at hand that has been shared by other authors (in the form of CMs). In this situation, the Query agent is used to obtain a first conceptualization of the subject, and the Formalizer agent is used to obtain the OWL formalization of that conceptualization. A portion of the OWL ontology obtained from the sample CM by the Formalizer agent is as follows:

xmlns:rdf="http://www.w3.org/1999/02/22-rdf-syntax-ns#"
xmlns:xsd="http://www.w3.org/2001/XMLSchema#"
xmlns:rdfs="http://www.w3.org/2000/01/rdf-schema#"
xmlns:owl="http://www.w3.org/2002/07/owl#"

```
xmlns="http://www.owl-ontologies.com/unnamed.owl#"
xml:base="http://www.owl-ontologies.com/unnamed.owl">
<owl:Ontology rdf:about="Water"/>
<owl:Class rdf:ID="Hydrogen"/>
<owl:Class rdf:ID="Oxygen"/>
<owl:Class rdf:ID="State"/>
<owl:Class rdf:ID="Air"/>
<owl:Class rdf:ID="Photosynthesis"/>
<owl:Class rdf:ID="Solar Energy"/>
<owl:Class rdf:ID="Solar Light"/>
<State rdf:ID="Liquid"/>
<owl:Class rdf:ID="Water">
    <rdfs:subClassOf>
        <owl:Restriction>
            <owl:onProperty rdf:resource="#state"/>
            <owl:hasValue rdf:resource="#Liquid">
        </owl:Restriction>
    </rdfs:subClassOf>
    <rdfs:subClassOf>
        <owl:Restriction>
            <owl:onProperty rdf:resource="#made of"/>
            <owl:someValuesFrom rdf:resource="#Hydrogen"/>
        </owl:Restriction>
    </rdfs:subClassOf>
    <rdfs:subClassOf>
        <owl:Restriction>
            <owl:onProperty rdf:resource="#made of"/>
            <owl:someValuesFrom rdf:resource="#Oxygen"/>
        </owl:Restriction>
    </rdfs:subClassOf>
</owl:Class>
<owl:Class rdf:ID="Mineral Water"/>
    <rdfs:subClassOf rdf:resource="#Water"/>
</owl:Class>
<owl:ObjectProperty rdf:about ="#hasPart">
    <rdfs:domain rdf:resource="#Air"/>
    <rdfs:range rdf:resource="#Oxygen"/>
</owl:ObjectProperty>
```

The theoretical modeling method presented has been applied to several concept maps about the environmental domain, with the aim of automatically creating the corresponding ontologies. These CMs have been constructed from different sources (texts and figures) with the assistance of environment experts.

4. Conclusions

The science underlying many environmental problems is getting clearer and Web makes everything more accessible. But the downside of clarity and accessibility is an increased obligation to respond. Biomonitcring and more sensitive measurement tools, for example, can now identify, at trace levels, virtually every chemical or emission found in the environment, whether in a polar bear in the Arctic Circle or in a wastewater treatment plant in Catalonia; and the data can be made almost-instantly available online.

The ripples of the Digital Age continue to move through our economy and society. The famous Moores Law predicts that the density of transistors on a microchip will double every 18 months. This trend has held for 40 years and relentlessly drives computing power up and the cost of digital technology down. For billions of people, an endless variety of information, and perhaps a degree of disinformation, is just a click away [9]. To make it possible for a more and more powerful Web to understand and satisfy the requests of people and for machines to use the Web content, some kind of semantic Web is necessary, an evolving extension of the World Wide Web in which the semantics of information and services is defined.

To realize a semantic Web, a universal medium for data, information and knowledge exchange, tools are required that allow users with little technical background to generate their own ontologies and collaborate in the construction of distributed knowledge bases. The work presented here is a contribution to the creation of these tools: a method to formally obtain ontologies codified in the OWL language from an informal knowledge representation, such as concept maps. In this article, we have combined the semantic analysis of linking-words in concept maps, mechanisms of natural language processing based on a concept-sense-disambiguation algorithm, and two knowledge bases (WordNet and a concept-map repository) in a novel way. Ultimately, this mapping between concept maps and OWL ontologies creates the bases for the collaborative development of ontologies in a more intuitive, friendlier manner for humans.

The integration of WordNet and the definition of a concept-sense-disambiguation algorithm, combined with the topological analysis of concept maps, allow maintaining a high flexibility during concept map construction. These aspects are important for less-expert users in ontology construction, and allow to augment the semantic inference in concept maps and to obtain more expressiveness in the resulting OWL. The method presented here is mainly applicable to shallow domains, due to the fact that the terms in WordNet are about general knowledge; and this is an important restriction to be taken into account. The use of domain ontologies as alternative knowledge sources (to be used in a way similar to WordNet) and the increase of the use of concept maps repositories will allow accelerating the applicability of this method to any kind of environmental domain, and should be considered as research lines for future work.

Inferred Sets	Basic Structure	OWL Coding

S_CS	(C, C')	`<owl:Class rdf:ID = " C "/>`
		`<owl:Class rdf:ID = " C' "/>`
		`<rdfs:subClassOf rdf:resource = " # C " />`
		`</owl:Class>`

S_CI	(C, I)	`<owl:Class rdf:ID = " C "/>` `< C rdf:ID=" I " />`

TABLE 2. Conventions for simple class and instance coding.

Inferred Sets	Basic Structure	OWL Coding

Intersection$_{classes}$	(C, {... C'$_i$ ...})	`<owl:Class rdf:ID = " C ">`
		`<rdfs:subClassOf>`
		`<owl:intersectionOf rdf:parseType="Collection">`
		`<owl:Class rdf:about = "# C'$_i$" />`
		`</owl:intersectionOf>`
		`</rdfs:subClassOf>`
		`</owl:Class>`

Intersection$_{class-property}$	(CI, C, Pr, V)	`<owl:Class rdf:ID = " CI ">`
		`<rdfs:subClassOf>`
		`<owl:intersectionOf rdf:parseType="Collection">`
		`<owl:Class rdf:about = "# C " />`
		`<owl:Restriction>`
		`<owl:onProperty rdf:resource= "# Pr"/>`
		`<owl:hasValue rdf:resource = "# V " />`
		`</owl:Restriction>`
		`</owl:intersectionOf>`
		`</rdfs:subClassOf>`
		`</owl:Class>`

Union	(CU, {... Ci ...})	`<owl:Class rdf:ID = " CU ">`
		`<rdfs:subClassOf>`
		`<owl:unionOf rdf:parseType="Collection">`
		`<owl:Class rdf:about = "# C$_i$ " />`
		`</owl:unionOf>`
		`</rdfs:subClassOf>`
		`</owl:Class>`

TABLE 3. Conventions for complex class coding.

Inferred Sets	Basic Structure	OWL Coding
S_CP	(C, Pr)	<owl:Class rdf:ID = " C "/> <owl:Class rdf:ID = " Pr "/> <owl:ObjectProperty rdf:ID = " has + Pr "> <rdfs:domain rdf:resource = "# C "/> <rdfs:range rdf:resource = "# Pr "/> </owl:ObjectProperty>
S_CPV	(C, Pr, C')	<owl:Class rdf ID = " C "> <rdfs:subClassOf> <owl:Restriction> <owl:onProperty rdf:resource= "# Pr " /> <owl:someValuesFrom rdf:resource = "# C' " /> </owl:Restriction> </rdfs:subClassOf> </owl:Class>

TABLE 4. Conventions for simple property coding.

Inferred Sets	Basic Structure	OWL Coding
S_CPV$_{hasValor}$	(C, Pr, C')	<owl:Class rdf:ID = " C "> <rdfs:subClassOf> <owl:Restriction> <owl:onProperty rdf:resource = "# Pr " /> <owl:hasValue rdf:resource = "# C' " /> </owl:Restriction> </rdfs:subClassOf> </owl:Class>
Pr$_{symmetric}$	{...Pr$_i$...}	For each Pr$_i$: <owl:SymmetricProperty rdf:ID= " Pr$_i$" > <rdf:type rdf:resource= "http://www.w3.org/2002/07/owl#ObjectProperty"/> </owl:SymmetricProperty>
Pr$_{functional}$	{...Pr$_i$...}	For each Pr$_i$: <owl:FunctionalProperty rdf:ID = " Pr$_i$ "> <rdf:type rdf:resource= "http://www.w3.org/2002/07/owl#ObjectProperty"/> </owl:FunctionalProperty>

TABLE 5. Conventions for property's restrictions and characteristics coding.

References

1. J. Atserias, B. Casas, E. Cornelles, M. González, L. Padró, and M. Padró, *Freeling 1.3: Syntactic and semantic services in an open-source NLP library*, Proceedings of the Fifth International Conference on Language Resources and Evaluation (LREC 2006), May 2006, Genova, Italia.

2. F. Baader, D. Calvanese, D. McGuinness, P. Patel-Schneider, and Nardi D., *The description logic handbook: Theory, implementation, and applications*, Cambridge University Press, 2003.

3. F. Baader and Nutt W., *Basic description logics*, pp. 47–100, 2003, In [2].

4. Tim Berners-Lee, James Hendler, and Ora Lassila, *The semantic web*, Scientific American (2001).

5. D. Borrajo, L. Castillo, and J. M. Corchado, *Current Topics in Artificial Intelligence*, Lecture Notes in Computer Science, vol. 4788, Springer, 1984.

6. V. Brilhante, G. Macedo, and Macedo S., *Heuristic transformation of well-constructed conceptual maps into owl preliminary domain ontologies*, Workshop on Ontologies and their Applications (WONTO'2006), October 2006, Brazil.

7. A. J. Cañas and M. Carvalho, *Concept maps and ai: an unlikely marriage?*, Proceedings of Simpósio Brasileiro de Informática Educativa (SBIE 2004), 2004, Manaus, Brazil, pp. 1–10.

8. A. J. Cañas, G. Hill, R. Carff, N. Suri, J. Lott, G. Gómez, T. C. Eskridge, M. Arroyo, and R. Carvajal, *Cmaptools: A knowledge modeling and sharing environment*, Proceedings of the First International Conference on Concept Mapping, vol. 1, Universidad Pública de Navarra, 2004, Pamplona, Spain, pp. 125–133.

9. Daniel C. Esty and Andrew S. Winston, *Gre en to gold*, Yale University Press, New Haven and London, 2006.

10. R. Gregorio, C. E. Rui, A. Freire, A. Freire, and C. E. Bonifácio, *Reading and environmental education*, Proceedings of the Second International Conference on Concept Mapping 2006 (CMC'06) (A. J. Cañas and J. D. Novak, eds.), vol. 2, Universidad de Costa Rica, 2006, San José, Costa Rica, pp. 208–212.

11. P. J. Hayes, T. C. Eskridge, R. Saavedra, T. Reichherzer, M. Mehrotra, and Bobrovnikoff D., *Collaborative knowledge capture in ontologies*, Proceedings of the Third International Conference on Knowledge Capture, Association for Computing Machinery, 2005, Alberta, Canada, pp. 99–106.

12. J. Heinze, *Applications of concept mapping to undergraduate general education science courses*, Proceedings of the First International Conference on Concept Mapping (CMC'04) (A. J. Cañas, J. D. Novak, and Gonzlez F. M., eds.), Universidad Pública de Navarra, 2004, Pamplona, Spain.

13. J. Heinze-Fry and F. Ludwig, *Cmaptools facilitates alignment of local curriculum with state standards: A case study*, Proceedings of the Second International Conference on Concept Mapping 2006 (CMC'06) (A. J. Cañas and J. D. Novak, eds.), vol. 1, Universidad de Costa Rica, 2006, San José, Costa Rica, pp. 71–78.

14. B. Henderson-Sellers and Giorgini P., *Agent-oriented methodologies*, IGI Publishing, 2005.

15. R. Johnsonbaugh, *Matemáticas discretas (4ta ed.)*, Prentice Hall, México, 1999.

16. G. Jones, A. Robertson, J. Forbes, and Hollier G., *The harpercollins dictionary of environmental science*, Harpercollins, 1992.

17. R. J. Juli and S. Himangshu, *Conceptualized pedagogical change: Evaluating the effectiveness of the eps model by using concept mapping to assess student conceptual change*, Proceedings of the Second International Conference on Concept Mapping 2006 (CMC'06) (A. J. Cañas and J. D. Novak, eds.), vol. 1, Universidad de Costa Rica, 2006, San José, Costa Rica, pp. 136–143.

18. K. Kharade and S. Thomas, *Looking for an alternative strategy for teaching and testing: An experiment with concept mapping in an inclusive science classroom*, Proceedings of the Second International Conference on Concept Mapping 2006 (CMC'06) (A. J. Cañas and J. D. Novak, eds.), vol. 1, Universidad de Costa Rica, 2006, San José, Costa Rica, pp. 367–374.

19. M. Kharatmal and G. Nagarjuna, *A proposal to refine concept mapping for effective science learning*, Proceedings of the Second International Conference on Concept Mapping 2006 (CMC'06) (A. J. Cañas and J. D. Novak, eds.), vol. 1, Universidad de Costa Rica, 2006, San José, Costa Rica, pp. 1–7.

20. M. Luck, P. McBurney, O. Shehory, and S. Willmott, *Agent technology: Computing as interaction (a roadmap for agent based computing)*, AgentLink, 2005.

21. O. Magntorn and G. Helldén, *Reading nature from a bottom-up perspective*, Proceedings of the Second International Conference on Concept Mapping 2006 (CMC'06) (A. J. Cañas and J. D. Novak, eds.), vol. 1, Universidad de Costa Rica, 2006, San José, Costa Rica, pp. 485–493.

22. M. Messerotti, *Virtual observatories and virtual grids: The interplay in fully exploiting solar-terrestrial data*, Proceedings of the 20th International CODATA Conference, 2006, Beijing, China.

23. G.A. Miller, R. Beckwidth, C. Fellbaum, D. Gross, and K. J. Miller, *Introduction to wordnet: An on-line lexical database*, International Journal of Lexicography (1990), no. 4, 235–244.

24. B. M. Moon, A. J. Pino, and C. A. Hedberg, *Studying transformation: The use of cmaptools in surveying the integration of intelligence and operations*, Proceedings of the Second International Conference on Concept Mapping, vol. 1, Universidad de Costa Rica, 2006, San José, Costa Rica, pp. 527–533.

25. J. D. Novak and A. J. Cañas, *The theory underlying concept maps and how to construct them*, Technical Report IHMC CmapTools 2006-01, Institute for Human and Machine Cognition, Pensacola Fl, 32502, USA, January 2006.

26. J. D. Novak and D. B. Gowin, *Learning how to learn*, Cambridge University Press, 1984.

27. Q. Numi and G. C. Low, *Comparison of ten agent-oriented methodologies*, pp. 341–367, 2005, In [14].

28. J. Pavón, J. J. Gómez-Sanz, and R. Fuentes, *The ingenias methodology and tools*, pp. 236–276, 2005, In[14].

29. R. Pérez and B. Bowen, *Cmaps: An useful tool for improving a national environment monitoring system design*, Proceedings of the Third International Conference on Concept Mapping 2008 (CMC'08) (A. J. Cañas, P. Reiska. M. Åhlberg, and J. D. Novak, eds.), 2008, Tallinn, Estonia and Helsinki, Finland.

30. M. Åhlberg, E. Lehmuskallio, and J. Lehmuskallio, *Naturegate, concept mapping and cmaptools: Creating global networks of servers for improved learning about, in and for nature, ecosystems, biodiversity, and sustainable development*, Proceedings of the Second International Conference on Concept Mapping 2006 (CMC'06) (A. J. Cañas and J. D. Novak, eds.), vol. 1, Universidad de Costa Rica, 2006, San José, Costa Rica, pp. 457–460.

31. S. Rebich and C. Gautier, *Concept mapping to reveal prior knowledge and conceptual change in a mock summit course on global climate change*, Journal of Geoscience Education (2005), no. 4, 355–365.

32. J. Rye and P. A. Rubba, *An exploration of the concept map as an interview tool to facilitate the externalization of students' understandings about global atmospheric change*, Journal of Research in Science Teaching (1998), no. 5, 421–546.

33. A. Simón-Cuevas, L. Ceccaroni, and A. Rosete-Suárez, *Generation of owl ontologies from concept maps in shallow domains*, pp. 259–267, 2007, In[5].

34. ———, *An Approach to Formal Modeling of Environmental Knowledge via Concept Maps and Ontologies*, Proceedings of the International Congress on Environmental Modeling and Software 2008 (iEMSs'08), 2008, Barcelona, Spain.

35. A. Simón-Cuevas, L. Ceccaroni, A. Rosete-Suárez, A. Suárez-Rodríguez, and M. De la Iglesia-Campos, *A concept sense disambiguation algorithm for concept maps*, Proceedings of the Third International Conference on Concept Mapping 2008 (CMC'08), 2008, Estonia and Finland.

36. A. Simón-Cuevas, V. Estrada, A. Rosete-Suárez, and V. Lara, *Gecosoft: Un entorno colaborativo para la gestión del conocimiento con mapas conceptuales*, Proceedings of the Second International Conference on Concept Mapping 2006 (CMC'06), vol. 2, Universidad de Costa Rica, 2006, San José, Costa Rica, pp. 114–117.

37. M. Smith, Ch. Welty, and McGuinness D., *Owl web ontology language guide*, W3C, February 2004.

38. M. Tanrere, *Environmental problem solving in learning chemistry for high school students*, Journal of Applied Science in Environmental Sanitation (2008), no. 1, 47–50.

39. K. Zak, *The assessment of pre-service teachers' knowledge of the environment using concept maps*, Proceedings of The Minnesota Association for Environmental Education Twelfth Annual Conference (K. Gilbertson and A. Murphy, eds.), 2005, Finland.

40. Y. Zhao, *The use of a constructivist teaching model in environmental science at beijing normal university*, The China Papers (2003), 78–83.

Luigi Ceccaroni
Departament de Llenguatges i Sistemes Informàtics (LSI), Universitat Politècnica de
Catalunya (UPC), Campus Nord, Edif. Omega
C. Jordi Girona, 1-3
08034 Barcelona
Spain
e-mail: luigi@lsi.upc.edu

Alfredo Simón-Cuevas
Centro de Estudios de Ingeniería de Sistemas (CEIS), Facultad de Ingeniería Informática
Instituto Superior Politécnico *José Antonio Echeverría*, 114, No. 11901, Marianao
C. Habana
Cuba
e-mail: asimon@ceis.cujae.edu.cu

Alejandro Rosete-Suárez
Centro de Estudios de Ingeniería de Sistemas (CEIS), Facultad de Ingeniería Informática
Instituto Superior Politécnico *José Antonio Echeverría*, 114, No. 11901, Marianao
C. Habana
Cuba
e-mail: rosete@ceis.cujae.edu.cu

Mailyn Moreno-Espino
Centro de Estudios de Ingeniería de Sistemas (CEIS), Facultad de Ingeniería Informática
Instituto Superior Politécnico *José Antonio Echeverría*, 114, No. 11901, Marianao
C. Habana
Cuba
e-mail: my@ceis.cujae.edu.cu

Whitestein Series in Software Agent Technologies and Autonomic Computing

Edited by
**Monique Calisti (Editor-in-Chief), Marius Walliser,
Stefan Brantschen, and Marc Herbstritt**

This series reports new developments in agent-based software technologies and agent-oriented software engineering methodologies, with particular emphasis on applications in the area of autonomic computing and communications.
The spectrum of the series includes research monographs, high quality notes resulting from research and industrial projects, outstanding Ph.D. theses, and the proceedings of carefully selected conferences. The series is targeted at promoting advanced research and facilitating know-how transfer to industrial use.

BIRKHÄUSER

■ **Gschwind, T.**, IBM Zürich, Switzerland / **Pautasso, C.**, University of Lugano, Switzerland (eds.)

Emerging Web Services Technology, Vol. II

The book contains papers presented at the Workshop on Emerging Web Service Technology (WEWST'07). It focuses among others on: Adoption of RESTful Web Services; Dynamic Web Service Discovery, the delivery of well defined QoS guarantees; and the performance evaluation of Web Services.

2008. 191 pages. Softcover.
ISBN 978-3-7643-8863-8

■ **Schumacher, M.**, IBIS, University of Applied Sciences Western Switzerland, Switzerland / **Helin, H.**, TelioSonera, Finland / **Schuldt, H.**, DBIS, University Basel, Switzerland (eds.)

CASCOM: Intelligent Service Coordination in the Semantic Web

The book presents the CASCOM architecture for service delivery and coordination in intelligent agent-based peer-to-peer environments. To its users, CASCOM makes easy and seamless access available to Semantic Web Services anytime, anywhere, and using any device.

2008. 390 pages. Softcover.

ISBN 978-3-7643-8574-3

■ **Ma, H.**, CWI, Amsterdam, The Netherlands, / **Leung H.F.**, Chinese University of Hong Kong, Hong Kong, China

Bidding Strategies in Agent-Based Continous Double Auctions

Negotiation of software agents in continuous double auctions is a central concern of online auctions. To prepare bids for auctions, agents have to apply some bidding strategy. This book presents a new bidding strategy and tools to enhance the performance of existing ones.

2008. 150 pages. Softcover.
ISBN 978-3-7643-8729-7

■ **Pěchouček, M.**, Czech Technical University, Prague, Czech Republic / **Thompson, S.G.**, ET. Labs, Suffolk, U.K. / **Voos, H.**, University of Applied Sciences, Ravensburg-Weingarten, Germany (eds.)

Defense Industry Applications of Autonomous Agents and Multi-Agent Systems

The book reviews the state of the art in the field of defense and security related applications based on Intelligent Agent technologies.

2007. 180 pages. Softcover.
ISBN 978-3-7643-8570-5

■ **Calisti, M.**, Whitestein Technologies AG, Zürich,

Switzerland / **van der Meer, S.**, Waterford Institute of Technology, Ireland / **Strassner, J.**, Motorola, Inc., Schaumburg, IL, USA (eds.)

Advanced Autonomic Networking and Communication

The book is a reference of state-of-the-art efforts in autonomic networking and communication.

2007. 200 pages. Softcover.
ISBN 978-3-7643-8568-2

■ **Annicchiarico, R.**, IRCCS, Rome, Italy / **Cortés, U.**, Universidad Malaga, Spain / **Urdiales, C.**, Polytècnica de Catalunya, Barcelona, Spain (eds.)

Agent Technology and e-Health

2007. 156 pages. Softcover.
ISBN 978-3-7643-8546-0

■ **Moreno, A.** University of Tarragona, Spain / **Pavón, J.**, University of Madrid, Spain (eds.)

Issues in Multi-Agent Systems The AgentCities.ES Experience

2007. 240 pages. Softcover.
ISBN 978-3-7643-8542-2

■ **Pautasso, C.**, IBM Zürich, Switzerland / **Bussler, C.**, Cisco Systems Inc., San Jose, USA (eds.)

Emerging Web Services Technology

2007. 182 pages. Softcover.
ISBN 978-3-7643-8447-0